STUDENT'S SOLU

Third Edition

BASIC COLLEGE MATHEMATICS
A TEXT/WORKBOOK

STUDENT'S SOLUTIONS MANUAL

Third Edition

BASIC COLLEGE MATHEMATICS
A TEXT/WORKBOOK

Miller
Salzman
Lial

Prepared with the assistance of
Brian Hayes
Triton College

HarperCollinsPublishers

Cover photo: M. Angelo/West Light

ISBN 0-673-46280-3

Copyright © 1991, 1987 HarperCollins Publishers, Inc.
All Rights Reserved.
Printed in the United States of America.

93 94 95   6 5 4 3

PREFACE

This book provides complete worked-out solutions for many of the exercises and test problems in Basic College Mathematics: A Text/Workbook, Third Edition, by Charles D. Miller, Stanley A. Salzman, and Margaret L. Lial. Solutions are included for every other odd-numbered section exercise according to the pattern 3, 7, 11, ... (including summary exercises and appendix exercises), and for all chapter review exercises, chapter tests, cumulative review exercises, and the final examination. Solutions for the remaining odd-numbered section exercises, according to the pattern 1, 5, 9, ..., may be found at the back of the textbook.

This book should be used as an aid as you work to master your course work. Try to solve the exercises that your instructor assigns before you refer to the solutions in this book. Then, if you have difficulty, read these solutions to guide you in solving the exercises. The solutions have been written to follow the methods presented in the textbook.

You may find that some solutions in a set are given in greater detail than others. Thus, if you cannot find an explanation for a difficulty that you encountered in one exercise, you may find the explanation in a solution for a similar exercise elsewhere in the set.

Solutions that require answer graphs will refer you to the answer section of your textbook; these graphs are not included in this book.

The following people have made valuable contributions to the completion of this Student's Solutions Manual: Beatrice Shaftman and Sharon Swiglo, editors; Judy Martinez, typist; and Carmen Eldersveld, proofreader.

CONTENTS

1 WHOLE NUMBERS

1.1	Reading and Writing Whole Numbers	1
1.2	Addition of Whole Numbers	1
1.3	Subtraction of Whole Numbers	3
1.4	Multiplication of Whole Numbers	5
1.5	Division of Whole Numbers	7
1.6	Long Division	9
1.7	Rounding Whole Numbers	10
1.8	Roots and Order of Operations	11
1.9	Solving Word Problems	13
	Chapter 1 Review Exercises	16
	Chapter 1 Test	27

2 MULTIPLYING AND DIVIDING FRACTIONS

2.1	Basics of Fractions	30
2.2	Mixed Numbers	30
2.3	Factors	31
2.4	Writing a Fraction in Lowest Terms	33
2.5	Multiplication of Fractions	33
2.6	Applications of Multiplication	34
2.7	Dividing Fractions	35
2.8	Multiplication and Division of Mixed Numbers	36
	Chapter 2 Review Exercises	37
	Chapter 2 Test	42

3 ADDING AND SUBTRACTING FRACTIONS

3.1	Adding and Subtracting Like Fractions	45
3.2	Least Common Multiples	45
3.3	Adding and Subtracting Unlike Fractions	47
3.4	Adding and Subtracting Mixed Numbers	48
3.5	Order Relations and the Order of Operations	49
	Chapter 3 Review Exercises	50
	Chapter 3 Test	57

Cumulative Review: Chapters 1–3   60

4 DECIMALS

4.1	Reading and Writing Decimals	66
4.2	Rounding Decimals	66
4.3	Addition of Decimals	67
4.4	Subtraction of Decimals	68
4.5	Multiplication of Decimals	70
4.6	Division of Decimals	71
4.7	Writing Fractions as Decimals	74
Chapter 4 Review Exercises		76
Chapter 4 Test		85

5 RATIO AND PROPORTION

5.1	Ratios	88
5.2	Rates	88
5.3	Proportions	89
5.4	Solving Proportions	90
5.5	Applications of Proportions	92
Chapter 5 Review Exercises		93
Chapter 5 Test		99

6 RATIONAL EXPRESSIONS

6.1	Basics of Percent	102
6.2	Percents and Fractions	102
6.3	The Percent Proportion	105
6.4	Identifying the Parts in a Percent Problem	106
6.5	Using Proportions to Solve Percent Problems	106
6.6	The Percent Equation	108
6.7	Applications of Percent	109
6.8	Simple Interest	111
Chapter 6 Review Exercises		112
Chapter 6 Test		121

Cumulative Review: Chapters 4–6   123

7 MEASUREMENT

7.1	The English System	130
7.2	Denominate Numbers	131
7.3	The Metric System – Length	132
7.4	The Metric System – Volume and Weight	132
7.5	Metric to English Conversions	133
	Chapter 7 Review Exercises	134
	Chapter 7 Test	137

8 GEOMETRY

8.1	Basic Geometric Terms	139
8.2	Angles and Their Relationships	139
8.3	Rectangles and Squares	140
8.4	Parallelograms and Trapezoids	141
8.5	Triangles	141
8.6	Circles	142
8.7	Volume	143
8.8	Pythagorean Formula	143
8.9	Similar Triangles	144
	Chapter 8 Review Exercises	145
	Chapter 8 Test	153

9 BASIC ALGEBRA

9.1	Signed Numbers	156
9.2	Addition and Subtraction of Signed Numbers	157
9.3	Multiplication and Division of Signed Numbers	158
9.4	Order of Operations	159
9.5	Evaluating Expressions and Formulas	161
9.6	Solving Equations	162
9.7	Solving Equations with Several Steps	165
9.8	Applications	167
	Chapter 9 Review Exercises	168
	Chapter 9 Test	176

Cumulative Review: Chapters 7–9   178

10 STATISTICS

10.1	Circle Graphs	186
10.2	Bar Graphs and Line Graphs	186
10.3	Histograms and Frequency Polygons	187
10.4	Mean, Median, and Mode	187
Chapter 10 Review Exercises		188
Chapter 10 Test		191

11 CONSUMER MATHEMATICS

11.1	Compound Interest	194
11.2	Consumer Credit – True Annual Interest	195
11.3	Buying a House	196
11.4	Life Insurance	197
Chapter 11 Review Exercises		197
Chapter 11 Test		200

FINAL EXAM 202

APPENDIX A: CALCULATORS 208

APPENDIX B: INDUCTIVE AND DEDUCTIVE REASONING 209

CHAPTER 1 WHOLE NUMBERS

Section 1.1

3. 18,015
   - 0 in the hundreds place
   - 1 in the ten thousands place

7. 71,105
   The digits in the thousands period are 71.
   The digits in the ones period are 105.

11. 9613 in words is: nine thousand, six hundred thirteen.   (Do not use "and thirteen".)

15. 75,756,665 in words is: seventy-five million, seven hundred fifty-six thousand, six hundred sixty-five.) (Do not use "and sixty-five".)

19. Seven hundred eighty-five thousand, two hundred twenty-three in digits is:
    785,223.

23. Thirteen thousand, one hundred twelve in digits is:
    13,112.

Section 1.2

3. 18
   + 61

    79

7. 158
   + 340

    498

11. Line up: 412
              234
            + 143

              789

15. Line up: 12,142
              43,201
            + 23,103

              78,446

19. Line up: 38,204
            + 91,020

             129,224

23. 65
    + 77

     142
   - 5 + 7 = 12   Write 2, carry 1
   - 1 + 6 + 7 = 14   Write 14

27. 78
    + 83

     161
   - 8 + 3 = 11   Write 1, carry 1
   - 1 + 7 + 8 = 16   Write 16

31. 906
    + 875

     1781
   - 6 + 5 = 11   Write 1, carry 1
   - 1 + 0 + 7 = 8   Write 8
   - 9 + 8 = 17   Write 17

35. 278
    + 135

     413
   - 8 + 5 = 13   Write 3, carry 1
   - 1 + 7 + 3 = 11   Write 1, carry 1
   - 1 + 2 + 1 = 4   Write 4

2   Chapter 1   Whole Numbers

39.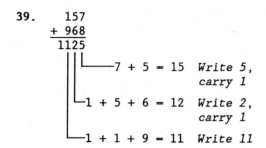

43.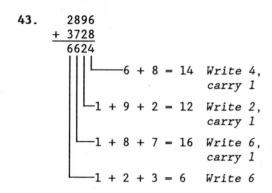

47.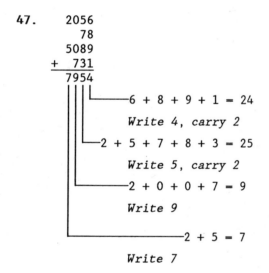

51.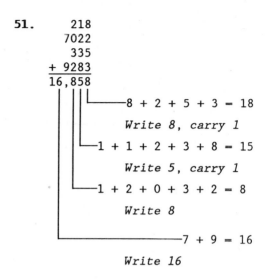

55.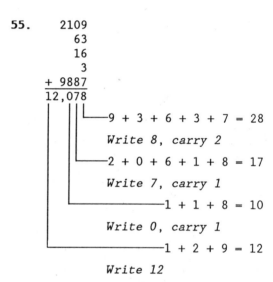

1.3 Subtraction of Whole Numbers   3

59.   413
         85
       9919
        602
         31
     + 1218
     12,268

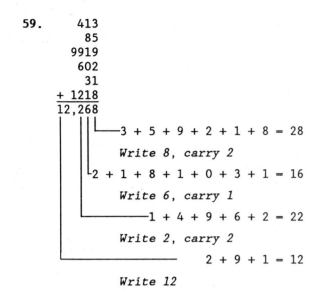

3 + 5 + 9 + 2 + 1 + 8 = 28
Write 8, carry 2
2 + 1 + 8 + 1 + 0 + 3 + 1 = 16
Write 6, carry 1
1 + 4 + 9 + 6 + 2 = 22
Write 2, carry 2
2 + 9 + 1 = 12
Write 12

63.   769   Add up to check the addition
       179
       214
       376
       759   Incorrect - should be 769

67.   11,577 Add up to check the addition
         678
        7952
          56
         718
        2173
       11,377 Incorrect - should be 11,577

71. Wilson and Thomasville

     Wilson to Southtown   11 miles
     Southtown to Thomasville   + 21 miles
                                   32 miles

     Any other route is longer.

75. Add the two costs to get the total cost.

         $19
       + $17
         $36

     The total cost is $36.

79. The church has: 1792
     The lodge has: + 3259
     Total books for sale: 5051

83.   308   Add all 3 sides together
       114   (It does not matter in
       286   what order you add the
       708 feet sides - remember the
             commutative property)

Section 1.3

3.   65   Check: 41
      - 24   + 24
        41   65

7.   79   Check: 62
      - 17   + 17
        62   79

11.   445   Check: 122
      - 323   + 323
       122   445

15.   777   Check: 665
      - 112   + 112
       665   777

19.   4420   Check: 4110
      - 310   + 310
      4110   4420

23.   7526   Check: 2213
     - 5313   + 5313
      2213   7526

27. 24,392   Check: 13,160
    - 11,232   + 11,232
     13,160   24,392

31.   24
      + 34
        58 Correct

Chapter 1 Whole Numbers

35.
```
  153
+ 131
  284  Incorrect;
```
Correct difference:
```
  274
- 153
  121
```

39.
```
  1421
+ 7212
  8633  Incorrect;
```
Correct difference:
```
  8643
- 1421
  7222
```

43.
```
   5 11
   6 1
 - 3 2
   2 9
```

47.
```
   5 11
   6 13
 - 2 51
   3 62
```

51.
```
   7 11
   58 1
 - 12 2
   45 9
```

55.
```
     17
     87 18
     99 8 8
 -   23 9 9
     75 8 9
```

59.
```
    17 12 12
    27  2  2 15
    38, 3  3  5
 -  29, 4  7  6
     8, 8  5  9
```

63.
```
    5 10
    6 0
  - 3 7
    2 3
```

67.
```
       9
    4 10 10
    5  0  0
 -  1  8  9
    3  1  1
```

71.
```
    5 10
    6 0 36
  - 5 8 22
      2 14
```

75.
```
    7 10
    158 0
  - 107 7
     50 3
```

79.
```
       9 11
    5 10 1 10
    6  0 2 0
 -  4  0 7 8
    1  9 4 2
```

83.
```
    7 10 6 9 15
    8  0, 7 0 5
 -  6  1, 6 6 7
    1  9, 0 3 8
```

87.
```
            9
    1 10 1 10 16
    2  0, 2  0  6
 -  1  8, 0  7  7
       2, 1  2  9
```

91.
```
  1628
+ 5954
  7582  Correct
```

95.
```
  17,346
+ 60,867
  78,213  Correct
```

99.
```
  14,552
+  2 553
  17,105  Incorrect;
```
Correct difference:
```
          9
       6 10 10
    17,  0  0 5
  - 14,  5  5 2
     2,  4  5 3
```

103.
```
   62 dogs
 - 40 dogs
   22 dogs remain
```

107.
```
   829 miles
 - 517 miles
   312 miles farther
```

111.
```
       11
    4  1 13
 $ 1  5  2  3  withheld
 -    1  3  7  9  owed
 $       1  4  4  refunded
```

115.
```
        12 14
    1  2  4 12
   12, 3  5  2  people on Tuesday
 - 11, 5  9  4  people on Monday
       7  5  8  more people on Tuesday
```

1.4 Multiplication of Whole Numbers

Section 1.4

3. $7 \times 1 \times 8$
    $= 7 \times 8$ $(7 \times 1 = 7)$
    $= 56$

7. $3 \cdot 1 \cdot 9$
    $= 3 \cdot 9$ $(3 \cdot 1 = 3)$
    $= 27$

11. 0   *Anything times 0 equals 0*

15. 28
    × 9
    ───
    252

    $8 \cdot 9 = 72$ *Write 2, carry 7*
    $9 \cdot 2 = 18$ *Add 7 to get 25, Write 25*

19. 624
    × 3
    ────
    1872

    $3 \cdot 4 = 12$ *Write 2, carry 1*
    $3 \cdot 2 = 6$ *Add 1 to get 7, Write 7*
    $3 \cdot 6 = 18$ *Write 18*

23. 2521
    × 4
    ─────
    10,084

    $4 \cdot 1 = 4$ *Write 4*
    $4 \cdot 2 = 8$ *Write 8*
    $4 \cdot 5 = 20$ *Write 0, carry 2*
    $4 \cdot 2 = 8$ *Add 2 to get 10, Write 10*

27. 21,835
    ×   6
    ────────
    131,010

    $6 \cdot 5 = 30$ *Write 0, carry 3*
    $6 \cdot 3 = 18$ *Add 3 to get 21, Write 1, carry 2*
    $6 \cdot 8 = 48$ *Add 2 to get 50, Write 0, carry 5*
    $6 \cdot 1 = 6$ *Add 5 to get 11, Write 1, carry 1*
    $6 \cdot 2 = 12$ *Add 1 to get 13, Write 13*

31. 20   2   20
    × 4   × 4   × 4
    ───   ───   ───
             8   80   *Attach 1 zero*

35. 740   74   740
    × 3   × 3   × 3
    ────   ───   ────
             222   2220 *Attach 1 zero*

39. 125   125   125
    × 30   × 3   × 30
    ────   ────   ────
            375   3750 *Attach 1 zero*

43. 800   8   800
    × 400   × 4   × 400
    ─────   ───────
                   320,000 *Attach 4 zeros*

47. $970 \cdot 50$ 97
                          × 5
                          ───
                          485

    $970 \cdot 50 = 48,500$

    Attach 2 zeros

51. $9700 \cdot 200$ 97
                           × 2
                           ───
                           194

    $9700 \cdot 200 = 1,940,000$

    Attach 4 zeros

Chapter 1 Whole Numbers

55.
```
      72
   ×  33
     216   ← 72 × 3
  + 216    ← 72 × 30 with 0 left off
    2376
```

59.
```
      58
   ×  41
      58
    232
    2378
```

63.
```
     758
   ×  24
    3 032
   15 16
   18,192
```

67.
```
     331
   ×  44
    1 324
   13 24
   14,564
```

71.
```
     638
   ×  555
    3190
    3 190
   31 90
   354,090
```

75.
```
    4355
  ×  615
   21 775
    4 355
   2 613 0
   2,678,325
```

79.
```
     135
   ×  401
     135
    0 00
   54 0
   54,135
```

83.
```
     837
   ×  708
    6 696
   00 00
   585 9
   592,596
```

87.
```
     3533
   × 5001
    3 533
   00 00
   000 0
   17 665
   17,668,533
```

91.
```
     3789
   × 2205
    18 945
    00 00
   757 8
   7 578
   8,354,745
```

95.
    12 cans per worker
 ×   9 workers
  108 Total cans

99.
    14 pictures
× $25 per picture
    70
    28
 $350 Total cost

103.
  108 boxes
× $37 per box
  756
  324
$3996 Total cost

107.
   36 miles
+ 104 miles
  140 Total miles

111.
  22 clerical employees
   9 management employees
  16 technical employees
+   8 warehouse employees
  55 total employees

Section 1.5

3. $9\overline{)45}^{\ 5}$ $45 \div 9 = 5$

7. $5 \div 5 = 1$ *Since* $5 \cdot 1 = 5$

11. $18 \div 0$ is meaningless.
    Dividing by zero is impossible

15. $12\overline{)0} = 0$ since $0 \cdot 12 = 0$. *Any number divided into zero is zero*

19. $\frac{0}{4} = 0$ since $4 \cdot 0 = 0$. *Any number divided into zero is zero*

23. $5\overline{)130}^{\ 26}$ Check: $\begin{array}{r} 26 \\ \times\ \ 5 \\ \hline 130 \end{array}$ Match

27. $3\overline{)1341}^{\ \ \ 447\ \ \ \ }_{\ \ \ \ 12}$ Check: $\begin{array}{r} 447 \\ \times\ \ \ 3 \\ \hline 1341 \end{array}$ Match

31. $4\overline{)2509}^{\ \ \ 627\ \text{R1}}_{\ \ \ \ 12}$ Check: $\begin{array}{r} 627 \\ \times\ \ \ 4 \\ \hline 2508 + 1 \\ 2509 \end{array}$ Match

35. $6\overline{)9137}^{\ 1522\ \text{R5}}$ Check: $\begin{array}{r} 1522 \\ \times\ \ \ 6 \\ \hline 9132 + 5 \\ 9137 \end{array}$ Match

39. $9\overline{)9054}^{\ 1006}$ Check: $\begin{array}{r} 1006 \\ \times\ \ \ 9 \\ \hline 9054 \end{array}$ Match

43. $6\overline{)4867}^{\ 811\ \text{R1}}$

    Check: $\begin{array}{r} 811 \\ \times\ \ \ 6 \\ \hline 4866 + 1 = 4867 \end{array}$ Match

47. $3\overline{)29{,}357}^{\ \ 9\ 785\ \ \text{R2}}_{\ \ \ \ \ 221}$

    Check: $\begin{array}{r} 9785 \\ \times\ \ \ \ 3 \\ \hline 29{,}355 + 2 = 29{,}357 \end{array}$ Match

51. $4\overline{)10{,}980}^{\ \ 2\ 745\ \ \ }_{\ \ \ \ \ 212}$

    Check: $\begin{array}{r} 2745 \\ \times\ \ \ \ 4 \\ \hline 10{,}980 \end{array}$ Match

55. $6\overline{)74{,}751}^{\ \ 12{,}458\ \text{R3}}_{\ \ \ \ \ 1\ 235}$

    Check: $\begin{array}{r} 12{,}458 \\ \times\ \ \ \ \ 6 \\ \hline 74{,}748 + 3 = 74{,}751 \end{array}$ Match

59. $7\overline{)71{,}776}^{\ \ 10{,}253\ \text{R5}}_{\ \ \ \ \ 132}$

    Check: $\begin{array}{r} 10{,}253 \\ \times\ \ \ \ \ 7 \\ \hline 71{,}771 + 5 = 71{,}776 \end{array}$ Match

63. $\begin{array}{r} 69 \\ \times\ \ 4 \\ \hline 276 + 1 = 277 \end{array}$ Correct

8   Chapter 1   Whole Numbers

67. 650
    $\underline{\times7}$
    $4550 + 2 = 4552$ Incorrect

    $7\overline{)4692}^{\,670\text{ R2}}$

    Check: 670
    $\underline{\times7}$
    $4690 + 2 = 4692$ Match

    Correct answer is 670 R2.

71. $^{652}20{,}763$
    $\underline{\times8}$
    $166{,}104$ Correct

75. $^{527}9628$
    $\underline{\times9}$
    $86{,}652 + 7 = 86{,}659$ Incorrect

    $9\overline{)86{,}655}^{\,9\,628\text{ R3}}$

    Check: 9628
    $\underline{\times9}$
    $86{,}655 + 3 = 86{,}652$ Match

    Correct answer is 9628 R3.

79. $^{1\,23}52136$
    $\underline{\times6}$
    $312{,}816 + 5 = 312{,}821$ Incorrect

    $6\overline{)312{,}820}^{\,52{,}136\text{ R4}}$

    Check: $^{1\,23}52136$
    $\underline{\times6}$
    $312{,}816 + 4 = 312{,}820$ Match

    Correct answer is 52,136 R4.

83. $7\overline{)\$57{,}764}^{\,\$\,8\,252}$ Each car costs $8252.

87. 30 is: • divisible (√) by 2, since it ends in 0.
    • divisible (√) by 3, since 3 + 0 = 3 which is divisible by 3.
    • divisible (√) by 5, since it ends in 0.
    • divisible (√) by 10, since it ends in 0.

91. 355 is: • not divisible (X) by 2, since it does not end in 0, 2, 4, 6, or 8.
    • not divisible (X) by 3, since 3 + 5 + 5 = 13 which is not divisible (X) by 3.
    • divisible (√) by 5, since it ends in 5.
    • not divisible (X) by 10, since it does not end in 0.

95. 2583 is: not divisible (X) by 2, since it does not end in in 2, 4, 6, 8 or 0.
    • divisible (√) by 3, since 2 + 5 + 8 + 3 = 18 which is divisible by 3.
    • not divisible (X) by 5, since it does not end in 0 or 5.
    • not divisible (X) by 10, since it does not end in 0.

99. The employee earns $16,200 in one year. Since there are 12 months in one year,

    $16{,}200 \div 12 = \$1350$

    are his earnings for each month.

1.6 Long Division

Therefore,
$$\$1350 \times 3 = \$4050$$
is the amount of earnings in three months.

Section 1.6

3.
```
      47 R10
53)2501
    212
    381
    371
     10
```

Check:
```
     47
   × 53
    141
   235
   2491 + 10 = 2501 Match
```

7.
```
     307 R4
23)7065
   69
   16
   00
   165
   161
     4
```

Check:
```
    307
  ×  23
    921
   614
   7061 + 4 = 7065 Match
```

11.
```
       274 R33
94)25,789
   18 8
    6 98
    6 58
      409
      376
       33
```

Check:
```
     274
   ×  94
    1 096
   24 66
   25,756 + 33 = 25,789 Match
```

15.
```
      1 239 R15
63)78,072
   63
   15 0
   12 6
    2 47
    1 89
      582
      567
       15
```

Check:
```
       1239
     ×   63
      3 717
     74 34
     78,057 + 15 = 78,072 Match
```

19.
```
      9 746 R1
12)116,953
   108
     8 9
     8 4
       55
       48
        73
        72
         1
```

Check:
```
       9746
     ×   12
      19,492
      97 46
     116,952 + 1 = 116,953 Match
```

23.
```
        3 331 R82
153)509,725
    459
     50 7
     45 9
      4 82
      4 59
        235
        153
         82
```

Check:
```
         3331
       ×  153
        9 993
       16,6 55
       33 3 1
       509,643 + 82 = 509,725 Match
```

27.
```
        1 114  R196
   657)732,094
       657
        75 0
        65 7
         9 39
         6 57
         2 824
         2 628
           196
```

Check:
```
      1114
    ×  657
     7 798
    55 70
   668 4
   731,898  + 196 = 732,094  Match
```

31.
```
      106
    ×  56
      636
     530
     5936 + 17 = 5953  Incorrect
```
```
        106 R7
    56)5943
       56
        34
        00
        343
        336
          7   Correct answer is 106 R7.
```

35.
```
       614
    ×   62
      1228
     3684
    36068 + 3 = 38,071  Incorrect
```
```
          62
    614)38068
        3684
        1228
        1228
           0   Correct answer is 62.
```

39.
```
    They must pay back    $3200
         plus interest  +  706
         amount repaid    $3906
```

43.
```
    42 circuits in one hour
   ×  8 hours in one day
   336 circuits in one day
```

```
    336 circuits in one day
   ×  5 days in a 5-day work week
   1680 circuits in a 5-day work week
```

Section 1.7

3. 127<u>6</u> rounded to the nearest ten is 1280.

7 is in tens position.

6 is 5 or larger.

7. 86,81<u>3</u> rounded to the nearest ten is 86,810

1 is in tens position.

3 is 4 or less.

11. 7<u>9</u>98 rounded to the nearest hundred is 8000.

9 is in hundreds position.

9 is 5 or larger.

(Note that when rounding the 9 up to the next larger hundred, 79 + 1 = 80.)

15. 31,0<u>5</u>2 rounded to the nearest hundred is 31,100.

0 is in hundreds position.

5 is 5 or larger.

19. 53,<u>1</u>82 rounded to the nearest thousand is 53,000.

3 is in thousands position.

1 is 4 or less.

1.8 Roots and Order of Operations   11

23. 8,9̲06,422 rounded to the nearest million is 9,000,000.
    8 is in millions position.
    9 is 5 or larger.

27. Ten: 4280   (428̲3; 3 is 4 or less)
    Hundred: 4300   (42̲83; 8 is 5 or more)
    Thousand: 4000   (4̲283; 2 is 4 or less)

31. Ten: 3130   (313̲2; 2 is 4 or less)
    Hundred: 3100   (31̲32; 3 is 4 or less)
    Thousand: 3000   (3̲132; 1 is 4 or less)

35. Ten:
    28,170   (28,171̲ is 4 or less)
    Hundred:
    28,200   (28,17̲1 is 5 or more)
    Thousand:
    28,000   (28,1̲71 is 4 or less)

39. 39   40   Each is
     14   10   rounded
     56   60   to the
    +73   +70   nearest ten
            180 ← is the estimated
                   answer.

43. 34   30   Each is rounded
    ×65   ×70   to the nearest ten
           2100 ← is the estimated
                   answer.

47. 614   600   Each is rounded to
    −276   −300   the nearest hundred
            300 ← is the estimated
                   answer.

51. Round 621,999,652 to the nearest
    Thousand: 622,000,000
        (621,999,6̲52; 6 is 5 or more)
    Ten thousand: 622,000,000
        (621,99̲9,652; 9 is 5 or more)
    Hundred thousand: 622,000,000
        (621,9̲99,652; 9 is 5 or more)

Section 1.8

3. 4, since $4 \cdot 4 = 16$.

7. 11, since $11 \cdot 11 = 121$.

11. Exponent is 2, base is 6;
    $6^2 = 6 \cdot 6 = 36$.

15. Exponent is 2, base is 15;
    $15^2 = 15 \cdot 15 = 225$.

19. $28^2 = 28 \cdot 28 = 784$ so $\sqrt{784} = 28$.

23. $40^2 = 40 \cdot 40 = 1600$ so $\sqrt{1600} = 40$.

27. $9^2 + 5 - 2$
    $= 81 + 5 - 2$ Exponents first
    $= 86 - 2$ Left to right
    $= 84$ Subtract last

31. $15 \cdot 2 \div 6$
    $= 30 \div 6$ Left to right
    $= 5$ Divide last

35. $6 \cdot 2^2 + \dfrac{0}{6}$
    $= 6 \cdot 4 + \dfrac{0}{6}$ Exponents first
    $= 24 + 0$ Multiply and divide
                     left to right
    $= 24$ Add last

39. $8 + 9 \div (5 - 2) + 6 \cdot 2$

   $= 8 + 9 \div 3 + 6 \cdot 2$ *Inside parentheses first*

   $= 8 + 3 + 12$ *Divide and multiply left to right*

   $= 23$ *Add last*

43. $6 \cdot \sqrt{144} - 8 \cdot 6$

   $= 6 \cdot 12 - 8 \cdot 6$ *Square root first*

   $= 72 - 48$ *Multiplication before subtraction*

   $= 24$ *Subtract last*

47. $6 + 10 \div 2 + 3 \cdot 3$

   $= 6 + 5 + 9$ *Divide and multiply left to right*

   $= 11 + 9$ *Add left to right*

   $= 20$ *Add last*

51. $3^2 + 6^2 + (30 - 21) \cdot 2$

   $= 3^2 + 6^2 + 9 \cdot 2$ *Inside parentheses*

   $= 9 + 36 + 9 \cdot 2$ *Simplify exponents*

   $= 9 + 36 + 18$ *Multiplication before addition*

   $= 45 + 18$ *Add left to right*

   $= 63$ *Add last*

55. $6 \cdot 3 - 8 \cdot 2 + 4$

   $= 18 - 8 \cdot 2 + 4$ *Multiply first (start at left)*

   $= 18 - 16 + 4$ *Other multiplication*

   $= 2 + 4$ *Add left to right*

   $= 6$ *Add last*

59. $6 \cdot (5 - 1) + \sqrt{4}$

   $= 6 \cdot 4 + \sqrt{4}$ *Inside parentheses*

   $= 6 \cdot 4 + 2$ *Square root*

   $= 24 + 2$ *Multiplication before addition*

   $= 26$ *Add last*

63. $5^2 \cdot 5^2 + (8 - 4)$

   $= 5^2 \cdot 2^2 + 4$ *Inside parentheses*

   $= 25 \cdot 4 + 4$ *Simplify exponents*

   $= 100 + 4$ *Multliplication before addition*

   $= 104$ *Add last*

67. $5 \cdot \sqrt{36} - 7 \cdot (7 - 4)$

   $= 5 \cdot \sqrt{36} - 7 \cdot 3$ *Inside parentheses*

   $= 5 \cdot 6 - 7 \cdot 3$ *Square root*

   $= 30 - 7 \cdot 3$ *Leftmost multiplication*

   $= 30 - 21$ *Other multiplication*

   $= 9$ *Subtract last*

71. $8 + 5 \div 5 + 7 \cdot 3$

   $= 8 + 1 + 7 \cdot 3$ *Divide and multiply left to right*

   $= 8 + 1 + 21$ *Multiplication before addition*

   $= 9 + 21$ *Add left to right*

   $= 30$ *Add last*

75. $3 \cdot \sqrt{25} - 4 \cdot \sqrt{9}$

   $= 3 \cdot 5 - 4 \cdot 3$ *Square roots*

   $= 15 - 4 \cdot 3$ *Leftmost multiplication*

   $= 15 - 12$ *Other multiplication*

   $= 3$ *Subtact last*

79. $15 \div 3 \cdot 2 \cdot 6 \div 3$ *Multiply and divide left to right*

$= 5 \cdot 2 \cdot 6 \div 3$ *Leftmost division*

$= 10 \cdot 6 \div 3$ *Leftmost multiplication*

$= 60 \div 3$ *Other multiplication*

$= 20$ *Other division*

83. $5 \div 1 \cdot 10 \cdot 4 \div (17 - 9)$

$= 5 \div 1 \cdot 10 \cdot 4 \div 8$ *Parentheses first*

$= 5 \cdot 10 \cdot 4 \div 8$ *Leftmost division*

$= 50 \cdot 4 \div 8$ *Leftmost multiplication*

$= 200 \div 8$ *Other multiplication*

$= 25$ *Other division*

87. $1 + 3 - 2 \cdot \sqrt{1} + 3 \cdot \sqrt{121} - 5 \cdot 3$

$= 1 + 3 - 2 \cdot 1 + 3 \cdot 11 - 5 \cdot 3$ *Square roots*

$= 1 + 3 - 2 + 3 \cdot 11 - 5 \cdot 3$ *Perform multiplication*

$= 1 + 3 - 2 + 33 - 5 \cdot 3$ *from left to right*

$= 1 + 3 - 2 + 33 - 15$

$= 4 - 2 + 33 - 15$ *Perform additions*

$= 2 + 33 - 15$ *and subtractions*

$= 35 - 15$ *from left to*

$= 20$ *right*

Section 1.9

3. Step 1: We need to find the total tickets sold in 7 days.

Step 2: We know how many tickets are sold each day and we know the total number of days.

Step 3: The total tickets sold should be about $400 \cdot 7$ or 2800.

Step 4: Multiply:

$$\begin{array}{r} 375 \\ \times\ \ 7 \\ \hline 2625 \end{array} \text{ tickets}$$

2625 is reasonably close to our estimate of 2800.

Check:

$$\begin{array}{r} 375 \\ 7\overline{)2625} \\ \underline{21}\ \ \\ 52\ \\ \underline{49}\ \\ 35 \end{array}$$

Answer: 2625 tickets

7. Step 1: We need to find the yards not used.

Step 2: We know the total yards and the number of yards used. We need to subtract.

$$\begin{array}{l} \text{total yards} \\ \underline{-\ \text{yards used}} \\ \text{yards not used} \end{array}$$

Step 3: Yards not used should be about 800 less than the total or $900 - 800 = 100$.

Step 4: Subtract:

$$\begin{array}{r} {\scriptstyle 8\ 13} \\ \not{9}\ \not{3}8 \\ -\ 7\ 83 \\ \hline 1\ 55 \end{array} \text{ yards}$$

155 yards is reasonably close to the estimate of 100 yards.

Check:

$$\begin{array}{r} {\scriptstyle 1}\ \ \\ 783 \\ +\ 155 \\ \hline 938 \end{array}$$

Answer: 155 yards

11. Step 1: We want to find the new bank balance.

Step 2: We know her old balance and how much she spent. We need to add up the total for tuition, books, and supplies and subtract it from her balance.
Old balance − money spent = new balance.

Step 3: New balance should be about 700 − (300 + 150 + 10) or 700 − 460 = 240.

Step 4: Add:
$$\begin{array}{r}\overset{11}{}\\286\\148\\12\\\hline 446\end{array}$$
Subtract:
$$\begin{array}{r}698\\-446\\\hline 252\end{array}$$

$252 is reasonably close to the estimate of $240.

Check:
$$\begin{array}{r}446\\+252\\\hline 698\end{array}$$

Answer: $252

15. Step 1: We need to find the total square feet.

Step 2: We know the square feet in an acre and the number of acres. We need to multiply.

$$\begin{array}{r}\text{square feet}\\\times\text{ acres}\\\hline\text{total square feet}\end{array}$$

Step 3: The total should be about 40,000 · 4 or 160,000.

Step 4: Multiply:
$$\begin{array}{r}\overset{122}{}\\43,560\\\times4\\\hline 174,240\end{array}$$

174,240 is close to 160,000 (Note that estimates are not as close when they involve multiplication.)

Check:
$$\begin{array}{r}43,560\\122\\4\overline{)174,240}\end{array}$$

Answer: 174,240 square feet

19. Step 1: We need to find the miles traveled per day.

Step 2: We know the total miles and the number of days. We need to divide.

$$\frac{\text{total miles}}{\text{number of days}}$$

Step: 3 The miles per day should be about 600 ÷ 10 or 60.

Step 4: Divide:
$$\begin{array}{r}63\\9\overline{)567}\\54\\\hline 27\end{array}$$

63 is close to our estimate of 60.

Check:
$$\begin{array}{r}\overset{2}{}\\63\\\times9\\\hline 567\end{array}$$

Answer: 63 miles

23. Step 1: We need to find the total number of deer.

Step 2: We know the numbers of deer in three separate sightings. We need to add to find the total.

1.9 Solving Word Problems

    Step 3:   The total number should be about $25 + 225 + 500 = 750$ deer.

    Step 4:   Add:
$$\begin{array}{r} 24 \\ 232 \\ +\ 512 \\ \hline 768 \end{array}$$

768 deer is reasonably close to our estimate of 750 deer.

Answer: 768 deer

27. Step 1:   We need to find the total cost of 25 shirts and 9 pairs of socks.

    Step 2:   We know the cost of 5 shirts is five times as much. Likewise, the cost of 9 pairs of socks will be three times as much as the cost of 3 pairs of socks (which we know).

    Step 3:   The total cost should be about $5 \cdot 20 + 3 \cdot 0$ or about 100 dollars.

    Step 4:   Calculate
$$5 \cdot 18 + 3 \cdot 4 = 90 + 12 = 102,$$
which is close to our estimate of 100.

Answer: $102

31. Step 1:   The number of machines on hand at the end of the month must be found.

    Step 2:   Do two steps.

    1st: The number of machines distributed must be subtracted from the number on hand.

$$\begin{array}{r} \text{number of machines on hand} \\ -\ \text{number of machines distributed} \\ \hline \text{number of machines remaining} \end{array}$$

    2nd: The number of machines returned must be added to the number of machines remaining to find the number of machines on hand at the end of the month.

$$\begin{array}{r} \text{number of machines remaining} \\ +\ \text{number of machines returned} \\ \hline \text{number of machines on hand} \end{array}$$

    Step 3:   The answer should be about $330 - 40 - 20 - 80 + 20 + 40 + 110$ or about 340.

    Step 4:   Compute.

    1st:
```
  325  on hand (beginning of
-  35          month)
  ---  distributed
  290
-  23  distributed
  ---
  267
-  76  distributed
  ---
  191  remaining after
       distribution
```

    2nd:
```
  191  remaining after
       distribution
+  15  returned
  ---
  206
+  38  returned
  ---
  244
+ 108  returned
  ---
  352  on hand at the end of
       the month
```

The answer is close to the estimate of 340.

Answer: 352 machines

16   Chapter 1   Whole Numbers

Chapter 1   Review Exercises

1. 7816
   Digit in the thousands period is 7.
   Digits in the ones period are 816.

2. 78,915
   Digits in the thousands period are 78.
   Digits in the ones period are 915.

3. 206,792
   Digits in the thousands period are 206.
   Digits in the ones period are 792.

4. 1,768,710,618
   Digit in the billions period is 1.
   Digits in the millions period are 768.
   Digits in the thousands period are 710.
   Digits in the ones period are 618.

5. 725 in words is:
   seven hundred twenty-five. (Do not use "and twenty-five".)

6. 17,615 in word is:
   seventeen thousand, six hundred fifteen. (Do not use "and fifteen".)

7. 62,500,005 in words is:
   sixty-two million, five hundred thousand, five.

8. $\underset{8}{\underline{\text{eight thousand}}}$, $\underset{120}{\underline{\text{one hundred twenty}}}$
   = 8120

9. $\underset{600}{\underline{\text{six hundred million}}}$, $\underset{015}{\underline{\text{fifteen thousand}}}$
   $\underset{759}{\underline{\text{seven hundred fifty-nine}}}$
   = 600,015,759

10.   74
    + 29
    ─────
     103

11.   43
    + 77
    ─────
     120

12.   778
    + 459
    ─────
     1237

13.   914
      3708
    +   34
    ──────
      4656
      │││└─ 4 + 8 + 4 = 16
      │││   Write 6, carry 1
      ││└── 1 + 1 + 0 + 3 = 5
      ││   Write 5
      │└─── 9 + 7 = 16
      │   Write 6, carry 1
      └──── 1 + 3 = 4
            Write 4

14.   8215
         9
    + 7433
    ──────
    15,657
      │││└─ 5 + 9 + 3 = 17
      │││   Write 7, carry 1
      ││└── 1 + 1 + 3 = 5
      ││   Write 5
      │└─── 2 + 4 = 6
      │   Write 6
      └──── 8 + 7 = 15
            Write 15

Chapter 1 Review Exercises 17

15. 1108   Add up and down to check.
      566
     7201
    + 304
     9179
     ┌─ 8 + 6 + 1 + 4 = 19
     │   Write 9, carry 1
     ├─ 1 + 0 + 6 + 0 + 0 = 7
     │   Write 7
     ├─ 1 + 5 + 2 + 3 = 11
     │   Write 1, carry 1
     └─ 1 + 1 + 7 = 9
         Write 9

16.   187   Add up and down to check.
     5543
      246
    + 1003
     6979
     ┌─ 7 + 3 + 6 + 3 = 19
     │   Write 9, carry 1
     ├─ 1 + 8 + 4 + 4 + 0 = 17
     │   Write 7, carry 1
     ├─ 1 + 1 + 5 + 2 + 0 = 9
     │   Write 9
     └─ 5 + 1 = 6
         Write 6

17.   5 732   Add up and down to check.
     11,069   Use groups of tens when
         37   possible.
      1 595
    + 22,169
     40,602
     ┌─ 2 + 9 + 7 + 5 + 9 = 32
     │   Write 2, carry 3
     ├─ 3 + 3 + 6 + 3 + 9 + 6 = 30
     │   Write 0, carry 3
     ├─ 3 + 7 + 0 + 5 + 1 = 16
     │   Write 6, carry 1
     └─ 1 + 1 + 2 = 4
         Write 4

18.   26   Check: 15
    − 15   + 11
     11   26

19.   79   Check: 57
    − 57   + 22
     22   79

20.   12
      1̶2̶ 18
      2̶3̶ 8̶   Check: 199
     − 19 9   + 39
        3 9   238

21. 3 13 7 10
    4̶ 3̶ 8̶ 0̶   Check: 577
    − 5 7 7   + 3803
       3 8 0 3   4380

22. 4 11 10 10
    5̶ 2̶ 1̶ 0̶   Check: 883
    −   8 8 3   + 4327
         4 3 2 7   5210

23.   10
       1̶0̶ 15
       2̶2̶ 2̶ 5̶   Check: 1198
    − 1 1 9 8   + 1017
         1 0 1 7   2215

24.   11 10
       1̶1̶ 0̶ 10
       2̶2̶ 1̶ 0̶   Check: 1986
    −   1 9 8 6   + 224
             2 2 4   2210

25.   16 9 14
         8 6̶ 1̶0̶
       9̶9̶,7̶ 0̶ 4̶   Check: 73,838
    − 7 3,8 3 8   + 25,866
       2 5,8 6 6   99,704

26.   8
    × 8
     64

27.   7
    × 0
      0

28. $\begin{array}{r} 4 \\ \times\ 6 \\ \hline 24 \end{array}$

29. $6 \times 9 = 54$

30. $6 \times 6 = 36$

31. $4 \times 9 = 36$

32. $7 \cdot 8 = 56$

33. $9 \cdot 9 = 81$

34. $2 \times 4 \times 6$
    $8 \times 6 \quad (2 \times 4 = 8)$
    48

35. $9 \times 1 \times 5$
    $9 \times 5 \quad (9 \times 1 = 9)$
    45

36. $3 \times 3 \times 8$
    $9 \times 8 \quad (3 \times 3 = 9)$
    72

37. $2 \times 2 \times 2$
    $4 \times 2 \quad (2 \times 2 = 4)$
    8

38. $8 \cdot 0 \cdot 6 = 0$ Anything times 0 equals 0

39. $9 \cdot 9 \cdot 1$
    $= 81 \cdot 1 \quad (9 \cdot 9 = 81)$
    $= 81$

40. $5 \cdot 1 \cdot 7$
    $= 5 \cdot 7 \quad (5 \cdot 1 = 5)$
    $= 35$

41. $7 \cdot 7 \cdot 0 = 0$ Anything times 0 equals 0

42. $\begin{array}{r} 65 \\ \times\ 2 \\ \hline 130 \end{array}$
    $5 \cdot 2 = 10$ Write 0, carry 1
    $2 \cdot 6 = 12 + 1 = 13$ Write 13

43. $\begin{array}{r} 92 \\ \times\ 7 \\ \hline 644 \end{array}$
    $7 \cdot 2 = 14$ Write 4, carry 1
    $7 \cdot 9 = 63 + 1 = 64$ Write 64

44. $\begin{array}{r} 24 \\ \times\ 3 \\ \hline 72 \end{array}$
    $3 \cdot 4 = 12$ Write 2, carry 1
    $3 \cdot 2 = 6 + 1 = 7$ Write 7

45. $\begin{array}{r} 89 \\ \times\ 1 \\ \hline 89 \end{array}$
    $1 \cdot 9 = 9$ Write 9
    $1 \cdot 8 = 8$ Write 8

46. $\begin{array}{r} 39 \\ \times\ 6 \\ \hline 234 \end{array}$
    $6 \cdot 9 = 54$ Write 4, carry 5
    $6 \cdot 3 = 18 + 5 = 23$ Write 23

47.
```
    781
  ×   7
   5467
```
└─ 7·1 = 7   Write 7
└─ 7·8 = 56   Write 6, carry 5
└─ 7·7 = 49 + 5 = 54
              Write 54

48.
```
    349
  ×   4
   1396
```
└─ 4·9 = 36   Write 6, carry 3
└─ 4·4 = 16 + 3 = 19
              Write 9, carry 1
└─ 4·3 = 12 + 1 + 13
              Write 13

49.
```
   9163
 ×    5
 45,815
```

50.
```
   7259
 ×    2
 14,518
```

51.
```
   4480
 ×    5
 22,400
```

52.
```
  93,105
 ×     5
 465,525
```

53.
```
  21,873
 ×     8
 174,984
```

54.
```
     22
  ×  15
    110
     22
    330
```

55.
```
     52
  ×  36
    312
    156
   1872
```

56.
```
     98
  ×  12
    196
     98
   1176
```

57.
```
    708
  ×  65
   3540
   4248
  46,020
```

58.
```
    655
  ×  21
    655
   1310
  13,755
```

59.
```
    392
  ×  77
   2 744
  27 44
  30,184
```

60.
```
    5032
  ×   48
  40 256
  201 28
  241 536
```

61.
```
    543
  ×  658
   4 344
  27 15
  325 8
  357,294
```

62.
```
     18 chairs
  × $32 per chair
     36
     54
   $576 total cost
```

63.
```
     24 balls
  × $13 per ball
     72
     24
   $312 total cost
```

64.
```
    278 tires
  × $48 per tire
   2 224
  11 12
  $13,344 total cost
```

65.
```
    168 welders masks
  ×  $9 per mask
   $1512 total cost
```

66.
```
    50      5      50
  ×  7    × 7    ×  5
             35    350   Attach 1 zero
```

67.
```
    380     38     380
  ×  80    × 8    × 80
             304   30,400   Attach 2 zeros
```

68.
$\begin{array}{r}752\\ \times\ 400\\\hline\end{array}$ $\begin{array}{r}752\\ \times\ \ \ 4\\\hline 3008\end{array}$ $\begin{array}{r}752\\ \times\ 400\\\hline 300,800\end{array}$
     Attach 2 zeros

69.
$\begin{array}{r}16,000\\ \times\ \ 8\,000\\\hline\end{array}$ $\begin{array}{r}16\\ \times\ 8\\\hline 128\end{array}$,000,000
     Attach 6 zeros

70.
$\begin{array}{r}43,000\\ \times\ \ 2\,100\\\hline\end{array}$ $\begin{array}{r}43\\ \times\ 21\\\hline 903\end{array}$ 90,300,000
     Attach 5 zeros

71.
$\begin{array}{r}30,200\\ \times\ 20,000\\\hline\end{array}$ $\begin{array}{r}302\\ \times\ \ 2\\\hline 604\end{array}$ 604,000,000 Attach 6 zeros

72. $6 \div 3 = 2$ Check: $2 \cdot 3 = 6$

73. $35 \div 7 = 5$ Check: $5 \cdot 7 = 35$

74. $42 \div 6 = 7$ Check: $7 \cdot 6 = 42$

75. $36 \div 4 = 9$ Check: $9 \cdot 4 = 36$

76. $\frac{72}{8} = 9$ $(9 \cdot 8 = 72)$

77. $\frac{27}{9} = 3$ $(3 \cdot 9 = 27)$

78. $\frac{54}{6} = 9$ $(9 \cdot 6 = 54)$

79. $\frac{0}{6} = 0$ $(0 \cdot 6 = 0)$

80. $\frac{125}{0}$ is meaningless

    Division by 0 is not possible

81. $\frac{0}{35} = 0$ $(0 \cdot 35 = 0)$

82. $8\overline{)648}$ quotient 81 Check: $\begin{array}{r}81\\ \times\ \ 8\\\hline 648\end{array}$

83. $5\overline{)180}$ quotient 36 Check: $\begin{array}{r}36\\ \times\ \ 5\\\hline 180\end{array}$

84. $9\overline{)56,259}$ quotient $6\,251$

85. $76\overline{)26,752}$ quotient 352
$\begin{array}{r}22\ 8\\\hline 3\ 95\\ 3\ 80\\\hline 152\\ 152\\\hline 0\end{array}$

86. $576 \div 6$ $6\overline{)576}$ quotient 96

87. $2704 \div 18$ $18\overline{)2704}$ quotient 150 R4
$\begin{array}{r}18\\\hline 90\\ 90\\\hline 04\\ 00\\\hline 4\end{array}$

Check: $\begin{array}{r}150\\ \times\ \ 18\\\hline 1200\\ 150\\\hline 2700\\ +\ \ \ 4\\\hline 2704\end{array}$

Chapter 1 Review Exercises   21

88. 15525 ÷ 125
$$
\begin{array}{r}
124 \text{ R}25 \\
125 \overline{\smash{)}15{,}525} \\
\underline{12\ 5} \\
3\ 02 \\
\underline{2\ 50} \\
525 \\
\underline{500} \\
25
\end{array}
$$

$$
\begin{array}{r}
\text{Check:}\quad 124 \\
\times\ 125 \\
\hline
620 \\
2\ 48 \\
12\ 4 \\
\hline
15{,}500 \\
+\ \ \ 25 \\
\hline
15{,}525
\end{array}
$$

89. 11<u>8</u> rounded to the nearest ten is 120.
1 is in the tens position.
8 is 5 or larger.

90. 16,7<u>0</u>1 rounded to the nearest hundred is 16,700.
7 is in the hundreds position.
0 is 4 or less.

91. 19,<u>7</u>21 rounded to the nearest thousand is 20,000.
9 is in the thousands position.
7 is 5 or larger.

92. 6<u>7</u>,485 rounded to the nearest ten thousand is 70,000.
6 is in the ten thousands position.
7 is 5 or larger.

93. Round 1496 to the nearest
Ten: 1500
(149<u>6</u>; 6 is 5 or more, 9 becomes 10)
Hundred: 1500
(14<u>9</u>6; 9 is 5 or more)

Thousand: 1000
(1<u>4</u>96; 4 is 4 or less)

94. Round 10,056 to the nearest
Ten: 10,060
(10,05<u>6</u>; 6 is 5 or more)
Hundred: 10,100
(10,0<u>5</u>6; 5 is 5 or more)
Thousand: 10,000
(10,<u>0</u>56; 0 is 4 or less)

95. Round 98,201 to the nearest
Ten: 98,200
(98,20<u>1</u>); 1 is 4 or less)
Hundred: 98,200
(98,2<u>0</u>1; 0 is 4 or less)
Thousand: 98,000
(98,<u>2</u>01; 2 is 4 or less)

96. Round 352,118 to the nearest
Ten: 352,120
(352,11<u>8</u>; 8 is 5 or more)
Hundred: 352,100
(352,1<u>1</u>8; 1 is 4 or less)
Thousand: 352,000
(352,<u>1</u>18; 1 is 4 or less)

97. $\sqrt{4} = 2$ (2 · 2 = 4)

98. $\sqrt{144} = 12$ (12 · 12 = 144)

99. $\sqrt{81} = 9$ (9 · 9 = 81)

100. $\sqrt{196} = 14$ (14 · 14 = 196)

101. Exponent is 2, base is 3.
$3^2 = 3 \cdot 3 = 9$

22 Chapter 1 Whole Numbers

102. Exponent is 3, base is 5.
$$5^3 = 5 \cdot 5 \cdot 5 = 125$$

103. Exponent is 5, base is 3.
$$3^5 = 3 \cdot 3 \cdot 3 \cdot 3 \cdot 3 = 243$$

104. Exponent is 5, base is 4.
$$4^5 = 4 \cdot 4 \cdot 4 \cdot 4 \cdot 4 = 1024$$

105. $9^2 - 9$
$= 81 - 9$ *Exponent first*
$= 72$ *Subtract*

106. $2 \cdot 7 + 6$
$= 14 + 6$ *Multiply first*
$= 20$ *Add last*

107. $2 \cdot 3^2 \div 2$
$= 2 \cdot 9 \div 2$ *Exponent first*
$= 18 \div 2$ *Multiply from left to right*
$= 9$ *Divide last*

108. $9 \div 1 \cdot 2 \cdot 2 \div (11 - 2)$
$= 9 \div 1 \cdot 2 \cdot 2 \div 9$ *Parentheses first*
$= 9 \cdot 2 \cdot 2 \div 9$ *Perform multiplications and divisions one at a time from left to right*
$= 18 \cdot 2 \div 9$
$= 36 \div 9$
$= 4$

109. $\sqrt{9} + 2 \cdot 3$
$= 3 + 2 \cdot 3$ *Square root first*
$= 3 + 6$ *Multiply*
$= 9$ *Add last*

110. $6 \cdot \sqrt{16} - 6 \cdot \sqrt{9}$
$= 6 \cdot 4 - 6 \cdot 3$ *Square roots*
$= 24 - 6 \cdot 3$ *Leftmost multiplication*
$= 24 - 18$ *Other multiplication*
$= 6$ *Subtract last*

111. Step 1: Find total cost.
Step 2: We know how many cash registers there are and how much each one costs.
Number of registers × cost of each = total cost.
Step 3: Approximate answer is $20 \times 1000 = \$20,000$.
Step 4: $\$1356 \times 15 = \$20,340$
Check answer.
$\$20,340 \div 15 = \$1356$

112. Step 1: Find total revolutions.
Step 2: We know the revolutions per minute and the number of minutes.
Number of revolutions × minutes = total revolutions.
Step 3: Approximate answer is $1000 \cdot 100 = 100,000$.
Step 4: $1400 \cdot 60$
$= 84,000$ revolutions
Check: $84,000 \div 60 = 1400$

113.
```
  120 words per minute
×  12 minutes
  240
 120
 1440 total words
```

114. 8000 nails per keg
   ×   40 kegs
   320,000 nails *Attach 4 zeros*

115. 2000 hours per home
   ×   12 homes
   24,000 hours *Attach 3 zeros*

116. 80 miles per hour
   × 5 hours
   400 total miles

117. 24 cans per case
   × 24 cases
   96
   48
   576 total cans

118. Step 1: Find total monthly collections.

   Step 2: We know the number of daily customers and the daily rate. We know the number of week-end-only customers and the rate.
   Number of customers × daily rate + number of customers × week-end rate = total collections.

   Step 3: Approximate answer
   = 60 · 6 + 20 · 3
   = 360 + 60 = 420.

   Step 4: 56 · 6 + 23 · 3
   = 336 + 69 = $405 total

119. Step 1: Find amount of refund.

   Step 2: We know how much he paid and how much he was required to pay (owed). We need to find the difference.
   Amount paid − amount owed = refund.

   Step 3: Approximate answer:
   1300 − 1000 = 300

   Step 4: $1279 − $1080 = $199 refund.
   Check: 1080 + 199 = 1279

120. Step 1: Find how many hours it takes to produce all the plates.

   Step 2: We know the total number of plates and we know how many are produced per hour.
   $\frac{\text{total number}}{\text{number per hour}}$ = total hours

   Step 3: Approximate answer:
   $\frac{30{,}000}{1000} = 30$

   Step 4: $\frac{30{,}888}{936} = 33$ total hours
   Check: 33 · 936 = 30,888

121. Step 1: Find out how many pounds of pork are needed.

   Step 2: We know the total number of cans and we must divide that total by 175 since each group of 175 cans requires 1 pound. The number of groups × 1 pound = total pounds.

Step 3: Approximate answer:

$\frac{9000}{200} = 45$, $45 \cdot 1 = 45$

Step 4: $\frac{8750}{175} = 50$, $50 \cdot 1 = 50$ pounds needed.

Check: 1 pound $\cdot$ 50 = 50 pounds, 50 $\cdot$ 175 cans = 8750 cans

122. Step 1: Find the new account balance.

Step 2: We know the check amount and the old balance.
Old balance − check amount = new balance.

Step 3: Approximate answer:
400 − 100 = 300

Step 4: $382 − $135 = $247
Check: 247 + 135 = 382

123. Step 1: Find the total acres fertilized.

Step 2: We know the total amount of fertilizer and how much each acre needs.
Total pounds ÷ pounds needed per acre = total acres.

Step 3: Approximate answer:
6000 ÷ 300 = 20

Step 4: 5750 ÷ 250 = 23 acres can be fertilized.
Check: 23 $\cdot$ 250 = 5750

124. Step 1: Find the amount of fencing to be built.

Step 2: We know the total needed and the amount completed.
Total needed − amount completed = amount to be built.

Step 3: Approximate answer:
3500 − 2000 = 1500

Step 4: 3415 − 1786 = 1629 feet of fencing still needed.

125.
64
$\times\ \ 5$
$\overline{320}$

$5 \cdot 4 = 20$ Write 0, carry 2
$5 \cdot 6 = 30 + 2 = 32$ Write 32

126. 81 ÷ 9 = 9   Check: 9 $\cdot$ 9 = 81

127. 179   Check: 64
$-\ \ 64$ $+115$
$\overline{115}$ $\overline{179}$

128. 8 × 5 = 40

129. 662
$+\ 379$
$\overline{1041}$

130. 72
$\times\ 29$
$\overline{648}$
144
$\overline{2088}$

131. $\begin{array}{r} 13\\ 0\,\cancel{3}10 \\ 38,\cancel{1}\cancel{4}\cancel{0} \\ -6\,078 \\ \hline 32,062 \end{array}$ Check: $\begin{array}{r} 6\,078 \\ +\,32,062 \\ \hline 38,140 \end{array}$

132. 21 ÷ 7 = 3

133. $\begin{array}{r} 6 \\ \times\,5 \\ \hline 30 \end{array}$

134. $\frac{42}{6} = 7$ (7 · 6 = 42)

135. $\begin{array}{r} 7\,218 \\ 3 \\ 18 \\ 1\,791 \\ 82,623 \\ +\,1\,982 \\ \hline 93,635 \end{array}$ Add up and down to check.

   8 + 3 + 8 + 1 + 3 + 2 = 25
   Write 5, carry 2

   2 + 1 + 1 + 9 + 2 + 8 = 23
   Write 3, carry 2

   2 + 2 + 7 + 6 + 9 = 26
   Write 6, carry 2

   2 + 7 + 1 + 2 + 1 = 13
   Write 3, carry 1

   1 + 8 = 9
   Write 9

136. $\begin{array}{r} 623 \\ \times\,9 \\ \hline 5607 \end{array}$

   9 · 3 = 27   Write 7, carry 2

   9 · 2 = 18 + 2 = 20
   Write 0, carry 2

   9 · 6 = 54 + 2 = 56
   Write 56

137. $\frac{2}{0}$ is meaningless.
   Division by 0 is not possible

138. $\begin{array}{r} 138 \\ 4\overline{)552} \end{array}$

139. 49,509 ÷ 9 $\begin{array}{r} 5\,501 \\ 9\overline{)49,509} \end{array}$

140. $\begin{array}{r} 8430 \\ \times\,128 \\ \hline 67\,440 \\ 168\,60 \\ 843\,0 \\ \hline 1,079,040 \end{array}$

141. $\frac{5}{1} = 5$ (5 · 1 = 5)

142. $\begin{array}{r} 108 \\ 34\overline{)3672} \end{array}$

143. $\begin{array}{r} 38,571 \\ \times3 \\ \hline 115,713 \end{array}$

144. 286,753 in words is:
   two hundred eight-six thousand,
   seven hundred fifty-three.

145. 72<u>4</u>5 rounded to the nearest hundred
   is 7200.
   2 is in the hundreds position.
   4 is 4 or less.

146. <u>5</u>00,196 rounded to the nearest
   million is 1,000,000.
   A zero is in millions position.
   5 is 5 or larger.

147. $\sqrt{36} = 6$ (6 · 6 = 36)

148. $\sqrt{100} = 10 \quad (10 \cdot 10 = 100)$

149. $\begin{array}{r} 56 \text{ skateboard} \\ \times\ \$38 \text{ per board} \\ \hline 448 \\ 168 \\ \hline \$2128 \text{ total cost} \end{array}$

150. $\begin{array}{r} 72 \text{ ice machines} \\ \times\ \$435 \text{ per machine} \\ \hline 360 \\ 2\ 16 \\ 28\ 8 \\ \hline \$31{,}320 \text{ total cost} \end{array}$

151. $\begin{array}{r} 185 \text{ bats} \\ \times\ \$12 \text{ per bat} \\ \hline 370 \\ 185 \\ \hline \$2220 \text{ total cost} \end{array}$

152. $\begin{array}{r} 607 \text{ boxes} \\ \times\ \$26 \text{ per box} \\ \hline 3\ 642 \\ 12\ 14 \\ \hline \$15{,}782 \text{ total cost} \end{array}$

153. $\begin{array}{r} 52 \text{ cards per deck} \\ \times\ 9 \text{ decks} \\ \hline 468 \text{ total cards} \end{array}$

154. Step 1: Find the total boxes sold.

  Step 2: We know the number of boxes each Brownie sells and how many Brownies there are. Number of Brownies × number of boxes each sells = total boxes sold.

  Step 3: Approximate answer: 10 · 1000 = 10,000

  Step 4: 20 · 500 = 10,000 boxes sold.
  Check: 10,000 ÷ 500 = 20

155. Step 1: Find the total hours.

  Step 2: We know the total days and the number of hours watched per day. Total days × hours each day = total hours.

  Step 3: Approximate answer: 2 · 400 · 5 = 4000

  Step 4: 2 · 365 = 730 total days
  730 · 5 = 3650 total hours

156. Step 1: Find the amount needed to be raised.

  Step 2: We know the total to be raised and the amount already raised. Total to be raised − amount already raised = amount needed.

  Step 3: Approximate answer: 85,000 − 35,000 = 50,000

  Step 4: $84,235 − $34,872 = $49,363 more to raise.

157. Step 1: Find the amount each would receive.

  Step 2: We know the amount to be divided among the 4 people. Amount ÷ 4 = amount each would receive.

  Step 3: Approximate anwer: 3000 ÷ 4 = 750

  Step 4: $3156 ÷ 4 = $789 each
  Check: 789 · 4 = 3156

158. Step 1: Find how many more miles per tank using super unleaded.

Step 2: We know the miles using super unleaded and the miles using regular unleaded. Miles using super unleaded − miles using regular unleaded = how many more using super unleaded.

Step 3: Approximate answer: 650 − 600 = 50

Step 4: 635 − 583 = 52 miles per tank

159. Step 1: Find out how much each family will pay.

Step 2: We know the total cost and the number of families to share it. Cost ÷ number of families = amount each will pay.

Step 3: Approximate answer: 28,000 ÷ 7 = 4000

Step 4: $26,950 ÷ 7 = $3850 each
Check: 3850 · 7 = 26,950

160. Step 1: Find the wage per hour.

Step 2: We know the total wages and the hours worked per day. Total wages ÷ hours = wage per hour.

Step 3: Approximate answer: 100 ÷ 10 = 10

Step 4: $108 ÷ 9 = $12 per hour

Chapter 1 Test

1. 3022 in words is:
   three thousand, twenty-two.

2. 52,008 in words is:
   fifty-two thousand, eight.

3. One hundred thirty-eight thousand, eight in digits is: 138,008.

4.
```
    729
     83
   9821
 + 6073
 16,706
```

5.
```
  17 063    Add up and down to check
       7
      12
   1 505
  93 710
 +   333
 112,630
```

6.
```
      9 9
    6 10 10 15
    7̸ 0̸ 0̸ 5̸        Check:  3660
  − 4 8 8 9              + 2542
    2 1 1 6                6202
```

7.
```
   11
   5̸1 10
   6̸2 0̸2        Check:  3660
   36 60                 2542
   25 42                 6202
```

8. 6 × 5 × 4
   30 × 4   (6 × 5 = 30)
   120

9.
```
      3000              Check:      3 000
   ×    49                     49)147,000
     27 000
    120 00
    147,000
```

10.
```
       75               Check:       75
    ×  18                      18)1350
      600
       75
     1350
```

11.
```
     7381               Check:     7 381
   ×  603                     603)4,450,743
        1
    22 143
    00 00
   442 86
   4,450,743
```

12.
```
        7 747           Check:      7747
    16)123,952                   ×    16
       112                        46 482
        11 9                      77 47
        11 2                     123,952
           75
           64
           112
           112
             0
```

13. $\frac{791}{0}$ is meaningless.

 Division by zero is not possible

14.
```
       458              Check:       458
    84)38,472                    ×    84
       33 6                       1 832
        4 87                     36 64
        4 20                     38,472
          672
          672
            0
```

15.
```
         170            Check:       170
     450)76,500                  ×   450
         45 0                        000
         31 50                      8 50
         31 50                     68 0
            00                     76,500
             0
```

16. 785<u>4</u> rounded to the nearest ten is 7850.

 5 is in tens position.

 4 is 4 or less.

17. 76,6<u>7</u>1 rounded to the nearest hundred is 76,700.

 6 is in hundreds position.

 7 is 5 or greater.

18. 45,<u>6</u>98 rounded to the nearest thousand is 46,000.

 5 is in thousands position.

 6 is 5 or greater.

19. $6^2 + 8 + 8$

 $= 36 + 8 + 7$ *Exponent*

 $= 51$ *Add*

20. $7\sqrt{64} - 14 \cdot 2$

 $= 7 \cdot 8 - 14 \cdot 2$ *Square root*

 $= 56 - 28$ *Multiply*

 $= 28$ *Subtract*

21. ($485 + $500 + $515 + $425) − $785

 = ($985 + $515 + $425) − $785

 = $1500 + $425) − $785

 = $1925 − $785

 = $1140

22. $\dfrac{35{,}424 \text{ autos}}{288 \text{ autos per day}} = 123$ days

23. $1108 - (\$690 + \$185 + \$68)$
    = $1108 - (\$875 + \$68)$
    = $1108 - \$943$
    = $165

24. ```
      873 production employees
       74 office and clerical workers
    +  22 management personnel
      969 total employees
    ```

25. ```
      118 self-cleaning ovens per hour
    ×   4 hours
      472 self-cleaning ovens

      139 standard ovens per hour
    ×   4 hours
      556 standard ovens

      472 self-cleaning ovens
    + 556 standard ovens
     1028 total ovens
    ```

CHAPTER 2 MULTIPLYING AND DIVIDING FRACTIONS

Section 2.1

3. $\frac{2}{3}$ There are 3 parts, and 2 are shaded

7. $\frac{3}{11}$ There are 11 coins, and 3 are quarters

11. $\frac{13}{71}$

15. Numerator: 9; denominator: 8

19. Proper fractions have numerator (top) smaller than denominator (bottom).
    They are: $\frac{3}{4}, \frac{9}{11}, \frac{7}{15}$.
    Improper fractions have numerator (top) larger than denominator (bottom).
    They are: $\frac{3}{2}, \frac{19}{18}$.

23. $2 \times 3 \times 3$
    $= 6 \times 3$ $(2 \times 3 = 6)$
    $= 18$

27. $21 \div 3 = 7$

Section 2.2

3. $3\frac{4}{5}$, $3 \cdot 5 = 15$, $15 + 4 = 19$, $\frac{19}{5}$

7. $3\frac{4}{9}$, $9 \cdot 3 = 27$, $27 + 4 = 31$, $\frac{31}{9}$

11. $6\frac{1}{3}$, $3 \cdot 6 = 18$, $18 + 1 = 19$, $\frac{19}{3}$

15. $10\frac{3}{4}$, $4 \cdot 10 = 40$, $40 + 3 = 43$, $\frac{43}{4}$

19. $8\frac{4}{5}$, $5 \cdot 8 = 40$, $40 + 4 = 44 = \frac{44}{5}$

23. $22\frac{8}{9}$, $9 \cdot 22 = 198$, $198 + 8 = 206$, $\frac{206}{9}$

27. $17\frac{14}{15}$, $15 \cdot 17 = 255$, $255 + 14 = 269$, $\frac{269}{15}$

31. $\frac{9}{2}$, $\quad 2\overline{)9}^{4\ R1}$, $4\frac{1}{2}$
    $\phantom{2\overline{)}}\underline{8}$
    $\phantom{2\overline{)}}1$

35. $\frac{14}{11}$, $\quad 11\overline{)14}^{1\ R3}$, $1\frac{3}{11}$
    $\phantom{11\overline{)}}\underline{11}$
    $\phantom{11\overline{)}}3$

39. $\frac{22}{5}$, $\quad 5\overline{)22}^{4\ R2}$, $4\frac{2}{5}$
    $\phantom{5\overline{)}}\underline{20}$
    $\phantom{5\overline{)}}2$

43. $\frac{58}{5}$, $\quad 5\overline{)58}^{11\ R3}$, $11\frac{3}{5}$
    $\phantom{5\overline{)}}\underline{5}$
    $\phantom{5\overline{)}}08$
    $\phantom{5\overline{)}}\underline{5}$
    $\phantom{5\overline{)}}3$

47. $\frac{50}{7}$, $\quad 7\overline{)50}^{7\ R1}$, $7\frac{1}{7}$
    $\phantom{7\overline{)}}\underline{49}$
    $\phantom{7\overline{)}}1$

2.3 Factors

51. $\frac{106}{5}$, $\begin{array}{r} 21 \text{ R1} \\ 5\overline{)106} \\ \underline{10} \\ 06 \\ \underline{5} \\ 1 \end{array}$, $21\frac{1}{5}$

55. $255\frac{1}{8}$, $255 \cdot 8 = 2040$

   $2040 + 1 = 2041$, $\frac{2041}{8}$

59. $\frac{2573}{15}$, $\begin{array}{r} 171 \text{ R8} \\ 15\overline{)2573} \\ \underline{15} \\ 107 \\ \underline{105} \\ 23 \\ \underline{15} \\ 8 \end{array}$ $171\frac{8}{15}$

63. $15 \cdot 4 + 6$

   $= 60 + 6$ *Multiplication before addition*

   $= 66$ *Add last*

Section 2.3

3. Factorizations of 8:

   $1 \cdot 8$, $2 \cdot 4$

   The factors of 8 are 1, 2, 4, 8.

7. Factorizations of 18:

   $1 \cdot 18$, $2 \cdot 9$, $3 \cdot 6$

   The factors of 18 are 1, 2, 3, 6, 9, 18.

11. Factorizations of 64:

   $1 \cdot 64$, $2 \cdot 32$, $4 \cdot 16$, $8 \cdot 8$
   Use the divisibility rules

   The factors of 64 are 1, 2, 4, 8, 16, 32, 64.

15. 3 is prime. It is divisible only by 3 and 1.

19. 17 is prime. Try dividing it by prime numbers. $17 \div 2 = 8$ R1, $17 \div 3 = 6$ R1, $17 \div 5 = 3$ R2, $17 \div 7 = 2$ R3, $17 \div 11 = 1$ R6

23. 25 is not prime. It is divisible by 5. 25 is composite.

27. 45 is not prime. It is divisible by 3 and 5. 45 is composite.

31.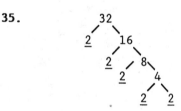

   $16 = 2 \cdot 2 \cdot 2 \cdot 2$ (4 factors of 2)
   Write this as 2^4.

35.

   $32 = 2 \cdot 2 \cdot 2 \cdot 2 \cdot 2$ (5 factors of 2)
   Write this as 2^5.

39.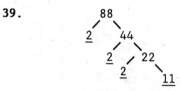

   $88 = 2 \cdot 2 \cdot 2 \cdot 11$ (3 factors of 2)
   Write this as $2^3 \cdot 11$.

43.

$100 = 2 \cdot 2 \cdot 5 \cdot 5$

(2 factors of 2 and 2 factors of 5)

Write this as $2^2 \cdot 5^2$.

47. 225 Use the divisibility rules.

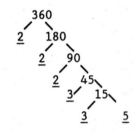

$225 = 3 \cdot 3 \cdot 5 \cdot 5$

(2 factors of 3 and 2 factors of 5)

Write this as $3^2 \cdot 5^2$.

51.

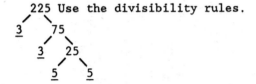

$360 = 2 \cdot 2 \cdot 2 \cdot 3 \cdot 3 \cdot 5$

(3 factors of 2 and 2 factors of 3)

Write this as $2^3 \cdot 3^2 \cdot 5$.

55. 8^3 (3 factors of 8)

$8^3 = 8 \cdot 8 \cdot 8$ 64
    $= 64 \cdot 8$ $\times\ 8$
    $= 512$ 512

59. $3^3 \cdot 4^2$ (3 factors of 3 and 2 factors of 4)

$3^3 \cdot 4^2 = 3 \cdot 3 \cdot 3 \cdot 4 \cdot 4$

    $= 27 \cdot 16$ 27
    $= 432$ $\times\ 16$
                       162
                       27
                      432

63. $6^2 \cdot 4^2$ (2 factors of 6 and 2 factors of 4)

$6^2 \cdot 4^2 = 6 \cdot 6 \cdot 4 \cdot 4$ 36
    $= 36 \cdot 16$ $\times\ 16$
    $= 576$ 216
                       36
                      576

67.

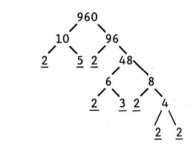

$960 = 2 \cdot 2 \cdot 2 \cdot 2 \cdot 2 \cdot 2 \cdot 3 \cdot 5$

(6 factors of 2)

Write this as $2^6 \cdot 3 \cdot 5$.

71. $1 \cdot 1 \cdot 5$

    $= 1 \cdot 5$ ($1 \cdot 1 = 1$)

    $= 5$

75. $36 \div 3 = 12$

Section 2.4

3. $\dfrac{20}{40} = \dfrac{20 \div 20}{40 \div 20} = \dfrac{1}{2}$

7. $\dfrac{36}{42} = \dfrac{36 \div 6}{42 \div 6} = \dfrac{6}{7}$

11. $\dfrac{180}{210} = \dfrac{180 \div 30}{210 \div 30} = \dfrac{6}{7}$

15. $\dfrac{12}{600} = \dfrac{12 \div 12}{600 \div 12} = \dfrac{1}{50}$

19. $\dfrac{10}{16} = \dfrac{\cancel{2} \cdot 5}{\cancel{2} \cdot 2 \cdot 2 \cdot 2} = \dfrac{1 \cdot 5}{1 \cdot 2 \cdot 2 \cdot 2} = \dfrac{5}{8}$

23. $\dfrac{60}{150} = \dfrac{\cancel{2} \cdot 2 \cdot \cancel{3} \cdot \cancel{5}}{\cancel{2} \cdot \cancel{3} \cdot 3 \cdot \cancel{5}} = \dfrac{1 \cdot 2 \cdot 1 \cdot 1}{1 \cdot 1 \cdot 1 \cdot 5} = \dfrac{2}{5}$

27. $\dfrac{77}{264} = \dfrac{7 \cdot \cancel{11}}{2 \cdot 2 \cdot 2 \cdot 3 \cdot \cancel{11}}$

    $= \dfrac{7 \cdot 1}{2 \cdot 2 \cdot 2 \cdot 3 \cdot 1} = \dfrac{7}{24}$

31. $\dfrac{12}{15} \times \dfrac{35}{45}$ cross products:

    $12 \cdot 45 = 540$

    $15 \cdot 35 = 525$

    The cross products are not equal, so the fractions are unequal.

35. $\dfrac{10}{25} \times \dfrac{16}{40}$ cross products

    $10 \cdot 40 = 400$

    $25 \cdot 16 = 400$

    The cross products are equal, so the fractions are equal.

39. $\dfrac{6}{50} \times \dfrac{9}{75}$ cross products:

    $6 \cdot 75 = 450$

    $50 \cdot 9 = 450$

    The cross products are equal, so the fractions are equal.

43. $\dfrac{232}{725} = \dfrac{2 \cdot 2 \cdot 2 \cdot \cancel{29}}{5 \cdot 5 \cdot \cancel{29}}$

    $= \dfrac{2 \cdot 2 \cdot 2 \cdot 1}{5 \cdot 5 \cdot 1} = \dfrac{8}{25}$

47. Factorizations of 64:

    $1 \cdot 64,\ 2 \cdot 32,\ 4 \cdot 16,\ 8 \cdot 8$

    Use the divisibility rules.

    The factors of 64 are

    1, 2, 4, 8, 16, 32, 64.

Section 2.5

3. $\dfrac{1}{9} \cdot \dfrac{1}{6} = \dfrac{1 \cdot 1}{9 \cdot 6} = \dfrac{1}{54}$

7. $\dfrac{\cancel{5}}{\cancel{6}} \cdot \dfrac{\cancel{12}}{25} \cdot \dfrac{3}{\cancel{4}} = \dfrac{1 \cdot 1 \cdot 3}{1 \cdot 5 \cdot 2} = \dfrac{3}{10}$

11. $\dfrac{9}{10} \cdot \dfrac{11}{16} = \dfrac{9 \cdot 11}{10 \cdot 16} = \dfrac{99}{100}$

    There are no common factors other than 1, so the answer is in the lowest terms.

Chapter 2 Multiplying and Dividing Fractions

15. $\dfrac{\cancel{15}^{3}}{7} \cdot \dfrac{8}{\cancel{25}_{5}} = \dfrac{3 \cdot 8}{7 \cdot 5} = \dfrac{24}{35}$

19. $10 \cdot \dfrac{3}{5} = \dfrac{\cancel{10}^{2}}{1} \cdot \dfrac{3}{\cancel{5}_{1}} = \dfrac{2 \cdot 3}{1 \cdot 1} = \dfrac{6}{1} = 6$

23. $30 \cdot \dfrac{1}{6} = \dfrac{\cancel{30}^{5}}{1} \cdot \dfrac{1}{\cancel{6}_{1}} = \dfrac{5 \cdot 1}{1 \cdot 1} = \dfrac{5}{1} = 5$

27. $100 \cdot \dfrac{21}{50} \cdot \dfrac{5}{14} = \dfrac{\cancel{100}^{\cancel{2}^{1}}}{1} \cdot \dfrac{\cancel{21}^{3}}{\cancel{50}} \cdot \dfrac{5}{\cancel{14}_{\cancel{2}_{1}}}$

$= \dfrac{1 \cdot 3 \cdot 5}{1 \cdot 1 \cdot 1} = \dfrac{15}{1} = 15$

31. $\dfrac{15}{32} \cdot 160 = \dfrac{15}{\cancel{32}} \cdot \dfrac{\cancel{160}^{5}}{1} = \dfrac{15 \cdot 5}{1 \cdot 1} = \dfrac{75}{1} = 75$

35. $\dfrac{54}{38} \cdot 684 \cdot \dfrac{5}{6} = \dfrac{\cancel{54}^{9}}{\cancel{38}_{1}} \cdot \dfrac{\cancel{684}^{18}}{1} \cdot \dfrac{5}{\cancel{6}_{1}}$

$= \dfrac{9 \cdot 18 \cdot 5}{1 \cdot 1 \cdot 1} = \dfrac{810}{1} = 810$

39. Multiply the length and the width.

$6 \cdot \dfrac{2}{3} = \dfrac{\cancel{6}^{2}}{1} \cdot \dfrac{2}{\cancel{3}_{1}} = \dfrac{4}{1} = 4$ square yards

43. Area = length · width

Area = $2 \cdot \dfrac{2}{3} = \dfrac{2}{1} \cdot \dfrac{2}{3} = \dfrac{4}{3}$

$= 1\dfrac{1}{3}$ square yards

Section 2.6

3. Multiply the 2 sides.

$\dfrac{1}{2} \cdot \dfrac{1}{2} = \dfrac{1 \cdot 1}{2 \cdot 2} = \dfrac{1}{4}$ square inch

7. Lupe earns

$\dfrac{3}{4}$ of $3600 in the summer.
↓ ↓   ↓

$\dfrac{3}{\cancel{4}_{1}} \cdot \dfrac{\cancel{3600}^{900}}{1} = \dfrac{3 \cdot 900}{1 \cdot 1} = \dfrac{2700}{1}$

= $2700 earned in the summer

11. $\dfrac{5}{11}$ of the volunteers are men.
↓ ↓

$\dfrac{5}{10} \cdot 165 = \dfrac{5}{\cancel{11}_{1}} \cdot \dfrac{\cancel{165}^{15}}{1} = \dfrac{5 \cdot 15}{1 \cdot 1}$

$= \dfrac{75}{1}$

= 75 are men

15. $\dfrac{1}{16}$ of $32,000

$= \dfrac{1}{\cancel{16}_{1}} \cdot \dfrac{\cancel{32{,}000}^{2000}}{1} = \dfrac{1 \cdot 2000}{1 \cdot 1}$

= $2000 saved

19. $\dfrac{5}{8}$ of 2400 votes

$= \dfrac{5}{\cancel{8}_{1}} \cdot \dfrac{\cancel{2400}^{300}}{1} = \dfrac{5 \cdot 300}{1 \cdot 1} = \dfrac{1500}{1}$

= 1500 votes needed from the south side

2.7 Dividing Fractions

23. Each carton contains 18 test kits. We must find out how many eighteens there are in 1332, and this is done by division.

$$18\overline{)1332}$$ quotient 74, with steps 126, 72, 72, 0.

74 cartons are needed.

Section 2.7

3. $\dfrac{5}{8} \div \dfrac{4}{3} = \dfrac{5}{8} \cdot \dfrac{3}{4} = \dfrac{5 \cdot 3}{8 \cdot 4} = \dfrac{15}{32}$

7. $\dfrac{7}{12} \div \dfrac{5}{18} = \dfrac{7}{\cancel{12}_2} \cdot \dfrac{\cancel{18}^3}{5} = \dfrac{7 \cdot 3}{2 \cdot 5} = \dfrac{21}{10} = 2\dfrac{1}{10}$

11. $\dfrac{\frac{36}{35}}{\frac{15}{14}} = \dfrac{36}{35} \div \dfrac{15}{14} = \dfrac{\cancel{36}^{12}}{\cancel{35}_5} \cdot \dfrac{\cancel{14}^2}{\cancel{15}_5} = \dfrac{12 \cdot 2}{5 \cdot 5} = \dfrac{24}{25}$

15. $15 \div \dfrac{3}{4} = \dfrac{15}{1} \div \dfrac{3}{4} = \dfrac{\cancel{15}^5}{1} \cdot \dfrac{4}{\cancel{3}_1} = \dfrac{5 \cdot 4}{1 \cdot 1}$

$= \dfrac{20}{1} = 20$

19. $\dfrac{8}{9}$ of an acre divided into 4 parts

$= \dfrac{8}{9} \div 4 = \dfrac{8}{9} \div \dfrac{4}{1} = \dfrac{\cancel{8}^2}{9} \cdot \dfrac{1}{\cancel{4}_1} = \dfrac{2 \cdot 1}{9 \cdot 1} = \dfrac{2}{9}.$

Each child will get 2/9 of an acre.

23. 7 ounces of medicine is to be divided into the 1/8 ounce vials.

$7 \div \dfrac{1}{8} = \dfrac{7}{1} \div \dfrac{1}{8} = \dfrac{7}{1} \cdot \dfrac{8}{1} = \dfrac{7 \cdot 8}{1 \cdot 1} = \dfrac{56}{1}$

56 of the vials can be filled.

27. The number of pages in the entire book can be found by dividing 320 by 5/8.

$\dfrac{320}{1} \div \dfrac{5}{8} = \dfrac{\cancel{320}^{64}}{1} \cdot \dfrac{8}{\cancel{5}_1} = \dfrac{64 \cdot 8}{1 \cdot 1} = \dfrac{512}{1}$

There are 512 pages in the entire book.

31. The total amount to be raised can be found by dividing 840,000 by 7/8.

$\dfrac{840{,}000}{1} \div \dfrac{7}{8} = \dfrac{\cancel{840{,}000}^{120{,}000}}{1} \cdot \dfrac{8}{\cancel{7}_1} = \dfrac{960{,}000}{1}$

The amount remaining to be raised is the total amount minus the amount raised so far.

$960{,}000 - 840{,}000 = 120{,}000$

$120,000 more must be raised.

35. $12\dfrac{2}{3}$, $12 \cdot 3 = 36$, $36 + 2 = 38$, $\dfrac{38}{3}$

Section 2.8

3. $1\frac{2}{3} \cdot 2\frac{7}{10} = \frac{5}{3} \cdot \frac{27}{10} = \frac{\cancel{5}}{\cancel{3}} \cdot \frac{\cancel{27}}{\cancel{10}}$

   $= \frac{1 \cdot 9}{1 \cdot 2} = \frac{9}{2} = 4\frac{1}{2}$

7. $10 \cdot 7\frac{1}{4} = \frac{10}{1} \cdot \frac{29}{4} = \frac{\cancel{10}}{1} \cdot \frac{29}{\cancel{4}} = \frac{5 \cdot 29}{1 \cdot 2}$

   $= \frac{145}{2} = 72\frac{1}{2}$

11. $9 \cdot 3\frac{1}{4} \cdot \frac{8}{3} = \frac{9}{1} \cdot \frac{13}{4} \cdot \frac{8}{3} = \frac{\cancel{9}}{1} \cdot \frac{13}{\cancel{4}} \cdot \frac{\cancel{8}}{\cancel{3}} = \frac{3 \cdot 13 \cdot 2}{1 \cdot 1 \cdot 1}$

    $= \frac{78}{1} = 78$

15. $2\frac{1}{2} \div 3\frac{3}{4} = \frac{5}{2} \div \frac{15}{4} = \frac{\cancel{5}}{\cancel{2}} \cdot \frac{\cancel{4}}{\cancel{15}} = \frac{1 \cdot 2}{1 \cdot 3} = \frac{2}{3}$

19. $3 \div 1\frac{1}{4} = \frac{3}{1} \div \frac{5}{4} = \frac{3}{1} \cdot \frac{4}{5} = \frac{3 \cdot 4}{1 \cdot 5} = \frac{12}{5}$

    $= 2\frac{2}{5}$

23. $5\frac{2}{3} \div \frac{1}{6} = \frac{17}{3} \div \frac{1}{6} = \frac{17}{\cancel{3}} \cdot \frac{\cancel{6}}{1} = \frac{17 \cdot 2}{1 \cdot 1}$

    $= \frac{34}{1} = 34$

27. $328\frac{1}{2}$ yards of gutter must be divided into sections consisting of $36\frac{1}{2}$ yards. $328\frac{1}{2} \div 36\frac{1}{2}$ will tell how many such sections there are.

$328\frac{1}{2} \div 36\frac{1}{2} = \frac{657}{2} \div \frac{73}{2} = \frac{\cancel{657}}{\cancel{2}} \cdot \frac{\cancel{2}}{\cancel{73}}$

$= \frac{9 \cdot 1}{1 \cdot 1} = \frac{9}{1} = 9$

9 homes can be filled with the rain gutter.

31. 11,875 pounds of paper are available, and each dictionary requires 2 3/8 pounds. 11,875 ÷ 2 3/8 will tell how many dictionaries can be published.

$11,875 \div 2\frac{3}{8} = \frac{11,875}{1} \div \frac{19}{8}$

$= \frac{\cancel{11,875}}{1} \cdot \frac{8}{\cancel{19}}$

$= \frac{625 \cdot 8}{1 \cdot 1}$

$= \frac{5000}{1}$

$= 5000$

5000 dictionaries can be published.

35. $\frac{6}{8} = \frac{\cancel{2} \cdot 3}{\cancel{2} \cdot 2 \cdot 2} = \frac{3}{4}$

39. $\frac{72}{80} = \frac{72 \div 8}{80 \div 8} = \frac{9}{10}$

Chapter 2 Review Exercises

1. $\frac{1}{4}$ There are 4 parts, and 1 is shaded

2. $\frac{5}{8}$ There are 8 parts, and 5 are shaded

3. $\frac{3}{4}$ There are 4 parts, and 3 are shaded

4. Proper fractions have numerator (top) smaller than denominator (bottom).
   They are: $\frac{3}{4}, \frac{2}{3}, \frac{1}{8}$.
   Improper fractions have numerator (top) larger than denominator (bottom).
   They are: $\frac{2}{1}, \frac{6}{5}$.

5. Proper fractions: $\frac{15}{16}, \frac{1}{8}$
   Improper fractions: $\frac{6}{5}, \frac{16}{13}, \frac{5}{3}$

6. $1\frac{1}{2}$, $2 \cdot 1 = 2$, $2 + 1 = 3$, $\frac{3}{2}$

7. $11\frac{5}{16}$, $16 \cdot 11 = 176$, $176 + 5 = 181$, $\frac{181}{16}$

8. $\frac{12}{5}$, $5\overline{)12}$ with quotient 2 R2, $2\frac{2}{5}$

9. $\frac{175}{13}$, $13\overline{)175}$ = 13 R6, $13\frac{6}{13}$

10. Factorizations of 6: $1 \cdot 6$, $2 \cdot 3$
    The factors of 6 are 1, 2, 3, 6.

11. Factorizations of 12: $1 \cdot 12$, $2 \cdot 6$, $3 \cdot 4$
    The factors of 12 are 1, 2, 3, 4, 6, 12.

12. Factorizations of 55: $1 \cdot 55$, $5 \cdot 11$
    The factors of 55 are 1, 5, 11, 55.

13. Factorizations of 90: $1 \cdot 90$, $2 \cdot 45$, $3 \cdot 30$, $5 \cdot 18$, $6 \cdot 15$, $9 \cdot 10$
    Use the divisibility rules
    The factors of 90 are 1, 2, 3, 5, 6, 9, 10, 15, 18, 30, 45, 90.

14.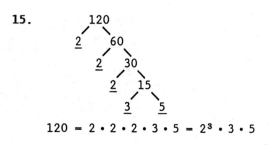
    $25 = 5 \cdot 5 = 5^2$

15.
    ```
        120
       /  \
      2   60
         /  \
        2   30
           /  \
          2   15
             /  \
            3    5
    ```
    $120 = 2 \cdot 2 \cdot 2 \cdot 3 \cdot 5 = 2^3 \cdot 3 \cdot 5$

38   Chapter 2   Multiplying and Dividing Fractions

16. *Use the divisibility rules*

$$225$$
$$3 \diagup \diagdown 75$$
$$3 \diagup \diagdown 25$$
$$5 5$$

$225 = 3 \cdot 3 \cdot 5 \cdot 5 = 3^2 \cdot 5^2$

17. $5^2 = 5 \cdot 5 = 25$

18. $2^2 \cdot 3^2 = 2 \cdot 2 \cdot 3 \cdot 3$
$ = 4 \cdot 9$
$ = 36$

19. $10^2 \cdot 3^3 = 10 \cdot 10 \cdot 3 \cdot 3 \cdot 3$
$ = 100 \cdot 27$
$ = 2700$

20. $3^5 \cdot 2^4 = 3 \cdot 3 \cdot 3 \cdot 3 \cdot 3 \cdot 2 \cdot 2 \cdot 2 \cdot 2$
$ = 243 \cdot 16$
$ = 3888$

21. $\dfrac{10}{15} = \dfrac{10 \div 5}{15 \div 5} = \dfrac{2}{3}$

22. $\dfrac{25}{35} = \dfrac{25 \div 5}{35 \div 5} = \dfrac{5}{7}$

23. $\dfrac{175}{190} = \dfrac{175 \div 5}{190 \div 5} = \dfrac{35}{38}$

24. $\dfrac{25}{60} = \dfrac{5 \cdot 5}{2 \cdot 2 \cdot 3 \cdot 5} = \dfrac{5}{12}$

25. $\dfrac{356}{480} = \dfrac{\cancel{2} \cdot \cancel{2} \cdot 89}{2 \cdot 2 \cdot \cancel{2} \cdot \cancel{2} \cdot 2 \cdot 3 \cdot 5}$
$\phantom{\dfrac{356}{480}} = \dfrac{1 \cdot 1 \cdot 89}{2 \cdot 2 \cdot 1 \cdot 1 \cdot 2 \cdot 3 \cdot 5}$
$\phantom{\dfrac{356}{480}} = \dfrac{89}{120}$

26. $\dfrac{3}{5} \diagdown\!\!\!\!\!\diagup \dfrac{15}{25}$ cross products:
$3 \cdot 25 = 75$
$5 \cdot 15 = 75$

The cross products are equal, so the fractions are equal.

27. $\dfrac{3}{8} \diagdown\!\!\!\!\!\diagup \dfrac{15}{40}$ cross products:
$3 \cdot 40 = 120$
$8 \cdot 15 = 120$

The cross products are equal, so the fractions are equal.

28. $\dfrac{1}{5} \cdot \dfrac{4}{5} = \dfrac{1 \cdot 4}{5 \cdot 5} = \dfrac{4}{25}$

29. $\dfrac{3}{5} \cdot \dfrac{3}{8} = \dfrac{3 \cdot 3}{5 \cdot 8} = \dfrac{9}{40}$

30. $\dfrac{\overset{5}{\cancel{70}}}{\underset{35}{\cancel{175}}} \cdot \dfrac{\overset{1}{\cancel{5}}}{\underset{7}{\cancel{14}}} = \dfrac{\overset{1}{\cancel{5}} \cdot 1}{\underset{7}{\cancel{35}} \cdot 1} = \dfrac{1 \cdot 1}{7 \cdot 1} = \dfrac{1}{7}$

31. $22 \cdot \dfrac{3}{11} = \dfrac{\overset{2}{\cancel{22}}}{1} \cdot \dfrac{3}{\underset{1}{\cancel{11}}} = \dfrac{2 \cdot 3}{1 \cdot 1} = \dfrac{6}{1} = 6$

32. $\dfrac{5}{16} \cdot 48 = \dfrac{5}{\underset{1}{\cancel{16}}} \cdot \dfrac{\overset{3}{\cancel{48}}}{1} = \dfrac{5 \cdot 3}{1 \cdot 1} = \dfrac{15}{1} = 15$

33. $\dfrac{5}{8} \cdot 1000 = \dfrac{5}{\underset{1}{\cancel{8}}} \cdot \dfrac{\overset{125}{\cancel{1000}}}{1}$
$\phantom{\dfrac{5}{8} \cdot 1000} = \dfrac{5 \cdot 125}{1 \cdot 1}$
$\phantom{\dfrac{5}{8} \cdot 1000} = \dfrac{625}{1}$
$\phantom{\dfrac{5}{8} \cdot 1000} = 625$

34. $\dfrac{2}{3} \cdot \dfrac{3}{4} = \dfrac{2}{3} \cdot \dfrac{4}{3} = \dfrac{2 \cdot 4}{3 \cdot 3} = \dfrac{8}{9}$

35. $\dfrac{5}{6} \div \dfrac{1}{2} = \dfrac{5}{\overset{}{\underset{3}{\cancel{6}}}} \cdot \dfrac{\overset{1}{\cancel{2}}}{1} = \dfrac{5 \cdot 1}{3 \cdot 1} = \dfrac{5}{3} = 1\dfrac{2}{3}$

36. $\dfrac{15}{18} \div \dfrac{10}{30} = \dfrac{\overset{5}{\cancel{15}}}{\underset{6}{\cancel{18}}} \cdot \dfrac{\overset{3}{\cancel{30}}}{\underset{1}{\cancel{10}}} = \dfrac{5 \cdot \overset{1}{\cancel{3}}}{\underset{2}{\cancel{6}} \cdot 1} = \dfrac{5 \cdot 1}{2 \cdot 1}$

 $= \dfrac{5}{2} = 2\dfrac{1}{2}$

37. $\dfrac{17}{20} \div \dfrac{34}{80} = \dfrac{\overset{1}{\cancel{17}}}{\underset{1}{\cancel{20}}} \cdot \dfrac{\overset{4}{\cancel{80}}}{\underset{2}{\cancel{34}}} = \dfrac{1 \cdot \overset{2}{\cancel{4}}}{1 \cdot \underset{1}{\cancel{2}}}$

 $= \dfrac{1 \cdot 2}{1 \cdot 1} = \dfrac{2}{1} = 2$

38. $10 \div \dfrac{5}{8} = \dfrac{10}{1} \div \dfrac{5}{8} = \dfrac{10}{1} \cdot \dfrac{8}{\underset{1}{\overset{2}{\cancel{5}}}}$

 $\dfrac{\overset{2}{\cancel{10}}}{1} \cdot \dfrac{8}{\underset{1}{\cancel{5}}} = \dfrac{2 \cdot 8}{1 \cdot 1} = \dfrac{16}{1} = 16$

39. $8 \div \dfrac{6}{7} = \dfrac{8}{1} \div \dfrac{6}{7} = \dfrac{8}{1} \cdot \dfrac{7}{\underset{3}{\overset{}{\cancel{6}}}}^{4} = \dfrac{4 \cdot 7}{1 \cdot 3} = \dfrac{28}{3} = 9\dfrac{1}{3}$

40. $1 \div \dfrac{7}{8} = \dfrac{1}{1} \div \dfrac{7}{8} = \dfrac{1}{1} \cdot \dfrac{8}{7} = \dfrac{1 \cdot 8}{1 \cdot 7} = \dfrac{8}{7} = 1\dfrac{1}{7}$

41. $\dfrac{2}{3} \div 5 = \dfrac{2}{3} \div \dfrac{5}{1} = \dfrac{2}{3} \cdot \dfrac{1}{5} = \dfrac{2 \cdot 1}{3 \cdot 5} = \dfrac{2}{15}$

42. $\dfrac{12}{13} \div 3 = \dfrac{12}{13} \div \dfrac{3}{1} = \dfrac{\overset{4}{\cancel{12}}}{13} \cdot \dfrac{1}{\underset{1}{\cancel{3}}} = \dfrac{4 \cdot 1}{13 \cdot 1} = \dfrac{4}{13}$

43. To find the area, multiply the length and the width.

 $\dfrac{7}{8} \cdot \dfrac{9}{16} = \dfrac{7 \cdot 9}{8 \cdot 16} = \dfrac{63}{128}$ square foot

44. To find the area, multiply the length and the width.

 $\dfrac{\overset{1}{\cancel{2}}}{3} \cdot \dfrac{7}{\underset{4}{\cancel{8}}} = \dfrac{1 \cdot 7}{3 \cdot 4} = \dfrac{7}{12}$ square inch

45. Multiply the length and the width.

 $12 \cdot \dfrac{1}{3} = \dfrac{\overset{4}{\cancel{12}}}{1} \cdot \dfrac{1}{\underset{1}{\cancel{3}}} = \dfrac{4 \cdot 1}{1 \cdot 1} = 4$ square yards

46. Multiply the length and the width.

 $40 \cdot \dfrac{5}{8} = \dfrac{\overset{5}{\cancel{40}}}{1} \cdot \dfrac{5}{\underset{1}{\cancel{8}}} = \dfrac{5 \cdot 5}{1 \cdot 1}$

 $= \dfrac{25}{1} = 25$ square feet

47. $3\dfrac{1}{2} \cdot 1\dfrac{1}{8} = \dfrac{7}{2} \cdot \dfrac{9}{8} = \dfrac{7 \cdot 9}{2 \cdot 8} = \dfrac{63}{16} = 3\dfrac{15}{16}$

48. $2\dfrac{1}{4} \cdot 7\dfrac{1}{8} \cdot 1\dfrac{1}{3} = \dfrac{\overset{3}{\cancel{9}}}{\underset{1}{\cancel{4}}} \cdot \dfrac{57}{8} \cdot \dfrac{\overset{1}{\cancel{4}}}{\underset{1}{\cancel{3}}}$

 $= \dfrac{3 \cdot 57 \cdot 1}{1 \cdot 8 \cdot 1}$

 $= \dfrac{171}{8}$

 $= 21\dfrac{3}{8}$

Chapter 2 Multiplying and Dividing Fractions

49. $12\frac{1}{3} \div 2 = \frac{37}{3} \div \frac{2}{1} = \frac{37}{3} \cdot \frac{1}{2}$

$= \frac{37 \cdot 1}{3 \cdot 2}$

$= \frac{37}{6} = 6\frac{1}{6}$

50. $3\frac{1}{8} \div 5\frac{5}{7} = \frac{25}{8} \div \frac{40}{7} = \frac{25}{8} \cdot \frac{7}{40}$ (with 25/40 reduced: 5/8)

$= \frac{5 \cdot 7}{8 \cdot 8}$

$= \frac{35}{64}$

51. 12 pounds is to be put into 3/4 pound cans.

$12 \div \frac{3}{4} = \frac{12}{1} \div \frac{3}{4} = \frac{12}{1} \cdot \frac{4}{3}$ (12/3 reduced to 4/1)

$= \frac{4 \cdot 4}{1 \cdot 1} = \frac{16}{1} = 16$

16 cans can be filled.

52. 2/3 of the estate is to be divided into 5 parts.

$\frac{2}{3} \div 5 = \frac{2}{3} \div \frac{5}{1} = \frac{2}{3} \cdot \frac{1}{5} = \frac{2 \cdot 1}{3 \cdot 5} = \frac{2}{15}$

Each child will receive 2/15 of the estate.

53. Find the number of times 4 3/8 will go into 78 3/4.

$78\frac{3}{4} \div 4\frac{3}{8} = \frac{315}{4} \div \frac{35}{8} = \frac{315}{4} \cdot \frac{8}{35}$ (315/35 = 9, 8/4 = 2)

$= \frac{9 \cdot 2}{1 \cdot 1} = \frac{18}{1} = 18$

18 blankets can be made.

54. Total wages equal the number of hours times the wages per hour.

$16\frac{1}{4} \cdot 6 = \frac{65}{4} \cdot \frac{6}{1} = \frac{65 \cdot 3}{2} = \frac{195}{2} = 97\frac{1}{2}$ (6/4 reduced to 3/2)

Mike earned 97 1/2 dollars, or $97.50.

55. The consumer sold 1/2 of 100

$\frac{1}{2} \cdot 100 = \frac{1}{2} \cdot \frac{100}{1} = \frac{1 \cdot 50}{1 \cdot 1}$

$= \frac{50}{1} = 50$ pounds.

100 − 50 = 50 pounds remaining.

She gave 2/5 of 50 to her parents.

$\frac{2}{5} \cdot 50 = \frac{2}{5} \cdot \frac{50}{1} = \frac{2 \cdot 10}{1 \cdot 1}$

$= \frac{20}{1} = 20$ pounds

50 − 20 = 30

The consumer has 30 pounds remaining.

56. Jennifer ran $\frac{2}{5}$ of $\frac{3}{4}$ mile.

$\frac{2}{5} \cdot \frac{3}{4} = \frac{2}{5} \cdot \frac{3}{4}$ (2/4 reduced to 1/2)

$= \frac{1 \cdot 3}{5 \cdot 2}$

$= \frac{3}{10}$

She ran 3/10 mile.

57. 5/8 must be divided by 4.

$$\frac{5}{8} \div 4 = \frac{5}{8} \div \frac{4}{1} = \frac{5}{8} \cdot \frac{1}{4} = \frac{5 \cdot 1}{8 \cdot 4} = \frac{5}{32}$$

Each employee will receive 5/32 of the profits.

58. $\dfrac{1}{\cancel{2}} \cdot \dfrac{\overset{1}{\cancel{2}}}{3} = \dfrac{1 \cdot 1}{1 \cdot 3} = \dfrac{1}{3}$

59. $\dfrac{7}{8} \div \dfrac{3}{4} = \dfrac{7}{\cancel{8}_{2}} \cdot \dfrac{\overset{1}{\cancel{4}}}{3} = \dfrac{7 \cdot 1}{2 \cdot 3} = \dfrac{7}{6} = 1\dfrac{1}{6}$

60. $12\dfrac{1}{2} \cdot 1\dfrac{2}{3} = \dfrac{25}{2} \cdot \dfrac{5}{3} = \dfrac{25 \cdot 5}{2 \cdot 3} = \dfrac{125}{6} = 20\dfrac{5}{6}$

61. $\dfrac{7}{8} \div 6 = \dfrac{7}{8} \div \dfrac{6}{1} = \dfrac{7}{8} \cdot \dfrac{1}{6} = \dfrac{7 \cdot 1}{8 \cdot 6} = \dfrac{7}{48}$

62. $\dfrac{\overset{1}{\cancel{42}}}{56} \cdot \dfrac{71}{\cancel{84}_{2}} = \dfrac{1 \cdot 71}{56 \cdot 2} = \dfrac{71}{112}$

63. $\dfrac{25}{31} \div \dfrac{50}{93} = \dfrac{\overset{1}{\cancel{25}}}{\cancel{31}_{1}} \cdot \dfrac{\overset{3}{\cancel{93}}}{\cancel{50}_{2}} = \dfrac{1 \cdot 3}{1 \cdot 2} = \dfrac{3}{2} = 1\dfrac{1}{2}$

64. $\dfrac{15}{31} \cdot 62 = \dfrac{15}{\cancel{31}_{1}} \cdot \dfrac{\overset{2}{\cancel{62}}}{1} = \dfrac{15 \cdot 2}{1 \cdot 1} = \dfrac{30}{1} = 30$

65. $3\dfrac{1}{4} \div 1\dfrac{1}{2} = \dfrac{13}{4} \div \dfrac{3}{2}$

$= \dfrac{13}{\cancel{4}_{2}} \cdot \dfrac{\overset{1}{\cancel{2}}}{3}$

$= \dfrac{13 \cdot 1}{2 \cdot 3}$

$= \dfrac{13}{6} = 2\dfrac{1}{6}$

66. $\dfrac{5}{2}$, $\;\;2\overline{)5}\;\dfrac{2\;\text{R}1}{}$, $\;2\dfrac{1}{2}$

67. $1\dfrac{6}{7}$, $\;7 \cdot 1 = 7,\; 7 + 6 = 13,\; \dfrac{13}{7}$

68. $\dfrac{198}{4}$, $\;\;4\overline{)198}\;\dfrac{49\;\text{R}2}{}$
$\;\;\;\;\;\;\;\;\;\;\;\;\;\;\;\underline{16}$
$\;\;\;\;\;\;\;\;\;\;\;\;\;\;\;\;38$
$\;\;\;\;\;\;\;\;\;\;\;\;\;\;\;\underline{36}$
$\;\;\;\;\;\;\;\;\;\;\;\;\;\;\;\;\;2$

$49\dfrac{2}{4} = 49\dfrac{1}{2}$

69. $12\dfrac{19}{21}$, $\;21 \cdot 12 = 252,\; 252 + 19 = 271$,

$\dfrac{271}{21}$

70. $\dfrac{8}{12} = \dfrac{\overset{1}{\cancel{2}} \cdot \overset{1}{\cancel{2}} \cdot 2}{\cancel{2} \cdot \cancel{2} \cdot 3} = \dfrac{1 \cdot 1 \cdot 2}{1 \cdot 1 \cdot 3} = \dfrac{2}{3}$

71. $\dfrac{108}{210} = \dfrac{\overset{1}{\cancel{2}} \cdot 2 \cdot \overset{1}{\cancel{3}} \cdot 3 \cdot 3}{\cancel{2} \cdot \cancel{3} \cdot 5 \cdot 7}$

$= \dfrac{1 \cdot 2 \cdot 1 \cdot 3 \cdot 3}{1 \cdot 1 \cdot 5 \cdot 7}$

$= \dfrac{18}{35}$

Chapter 2 Multiplying and Dividing Fractions

72. $\frac{10}{30} = \frac{10 \div 10}{30 \div 10} = \frac{1}{3}$

73. $\frac{75}{200} = \frac{75 \div 25}{200 \div 25} = \frac{3}{8}$

74. $\frac{29}{87} = \frac{29 \div 29}{87 \div 29} = \frac{1}{3}$

75. Each of the 43 5/9 yards need 3 1/2 ounces. (Think of whole numbers. If we had 4 yards and each needed 2 ounces, 4 · 2 = 8 ounces needed.)

    $43\frac{5}{9} \cdot 3\frac{1}{2} = \frac{\cancel{392}^{196}}{9} \cdot \frac{7}{\cancel{2}_1} = \frac{196 \cdot 7}{9 \cdot 1} = \frac{1372}{9}$

    $9)\overline{1372}$ 152 R4   or 152 4/9 ounces of glue needed

76. This is a difficult problem, so once again think of whole numbers. If a tank contains 10 gallons when it is 1/2 full, it will contain 20 gallons when it is full.

    $10 \div \frac{1}{2} = \frac{10}{1} \div \frac{1}{2} = \frac{10}{1} \cdot \frac{2}{1} = \frac{20}{1} = 20$

    Apply this method to the problem.

    $35 \div \frac{5}{8} = \frac{35}{1} \div \frac{5}{8} = \frac{\cancel{35}^{7}}{1} \cdot \frac{8}{\cancel{5}_1} = \frac{7 \cdot 8}{1 \cdot 1} = \frac{56}{1} = 56$

    The tank will hold 56 gallons when full.

77. 84 rolls of carpet are available, and each office requires 7/8 roll.
    84 ÷ 7/8 will tell how many offices can be carpeted.

    $84 \div \frac{7}{8} = \frac{84}{1} \div \frac{7}{8} = \frac{\cancel{84}^{12}}{1} \cdot \frac{8}{\cancel{7}_1}$

    $= \frac{12 \cdot 8}{1 \cdot 1} = \frac{96}{1} = 96$

    96 offices can be carpeted.

78. To find the area multiply the length and the width.

    $\frac{\cancel{2}^1}{\cancel{3}_1} \cdot \frac{\cancel{3}^1}{\cancel{4}_2} = \frac{1 \cdot 1}{1 \cdot 2} = \frac{1}{2}$ square inch

Chapter 2 Test

1. $\frac{3}{8}$ There are 8 parts and 3 are shaded

2. $\frac{1}{6}$ There are 6 parts and 1 is shaded

3. Proper fractions have numerator (top) smaller than denominator (bottom). They are: $\frac{5}{8}, \frac{7}{16}, \frac{2}{3}, \frac{3}{14}$.

4. $3\frac{5}{8}$, 8 · 3 = 24, 24 + 5 = 29, $\frac{29}{8}$

5. $\frac{143}{8}$, $8)\overline{143}$ 17 R7 , $17\frac{7}{8}$
   $\quad\quad\quad\quad\;\; \underline{8}$
   $\quad\quad\quad\quad\;\; 63$
   $\quad\quad\quad\quad\;\; \underline{56}$
   $\quad\quad\quad\quad\;\;\; 7$

6. Factorizations of 15:
   $1 \cdot 15, \; 3 \cdot 5$
   The factors of 15 are 1, 3, 5, and 15.

7.
   $40 = 2 \cdot 2 \cdot 2 \cdot 5 = 2^3 \cdot 5$

8.
   $75 = 3 \cdot 5 \cdot 5 = 3 \cdot 5^2$

9.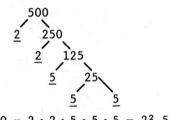
   $500 = 2 \cdot 2 \cdot 5 \cdot 5 \cdot 5 = 2^2 \cdot 5^3$

10. $\dfrac{12}{15} = \dfrac{2 \cdot 2 \cdot \cancel{3}}{\cancel{3} \cdot 5} = \dfrac{2 \cdot 2 \cdot 1}{1 \cdot 5} = \dfrac{4}{5}$

11. $\dfrac{56}{84} = \dfrac{\cancel{2} \cdot \cancel{2} \cdot 2 \cdot \cancel{7}}{\cancel{2} \cdot \cancel{2} \cdot 3 \cdot \cancel{7}} = \dfrac{1 \cdot 1 \cdot 2 \cdot 1}{1 \cdot 1 \cdot 3 \cdot 1} = \dfrac{2}{3}$

12. $\dfrac{5}{6} \cdot \dfrac{1}{3} = \dfrac{5 \cdot 1}{6 \cdot 3} = \dfrac{5}{18}$

13. $\dfrac{\cancel{6}^{\,2}}{7} \cdot \dfrac{2}{\cancel{9}_{\,3}} = \dfrac{2 \cdot 2}{7 \cdot 3} = \dfrac{4}{21}$

14. $6 \cdot \dfrac{4}{5} = \dfrac{6}{1} \cdot \dfrac{4}{5} = \dfrac{6 \cdot 4}{1 \cdot 5} = \dfrac{24}{5} = 4\dfrac{4}{5}$

15. $24 \cdot \dfrac{3}{4} = \dfrac{\cancel{24}^{\,6}}{1} \cdot \dfrac{3}{\cancel{4}_{\,1}} = \dfrac{6 \cdot 3}{1 \cdot 1} = \dfrac{18}{1} = 18$

16. Multiply the length and the width.
    $\dfrac{\cancel{4}^{\,1}}{\cancel{5}_{\,1}} \cdot \dfrac{\cancel{5}^{\,1}}{\cancel{16}_{\,4}} = \dfrac{1 \cdot 1}{1 \cdot 4} = \dfrac{1}{4}$ square inch

17. $\dfrac{7}{10}$ of 230 are drinking coffee.
    $\dfrac{7}{10} \cdot 230 = \dfrac{7}{\cancel{10}_{\,1}} \cdot \dfrac{\cancel{230}^{\,23}}{1}$
    $= \dfrac{7 \cdot 23}{1 \cdot 1}$
    $= \dfrac{161}{1} = 161$

    161 students are drinking coffee.

18. $\dfrac{3}{5} \div \dfrac{8}{10} = \dfrac{3}{\cancel{5}_{\,1}} \cdot \dfrac{\cancel{10}^{\,2}}{8} = \dfrac{3 \cdot \cancel{2}^{\,1}}{1 \cdot \cancel{8}_{\,4}} = \dfrac{3 \cdot 1}{1 \cdot 4} = \dfrac{3}{4}$

19. $\dfrac{7}{\tfrac{4}{9}} = 7 \div \dfrac{4}{9} = \dfrac{7}{1} \div \dfrac{4}{9} = \dfrac{7}{1} \cdot \dfrac{9}{4}$
    $= \dfrac{7 \cdot 9}{1 \cdot 4} = \dfrac{63}{4} = 15\dfrac{3}{4}$

Chapter 2 Multiplying and Dividing Fractions

20. How many times will 3/5 go into 75?

$$75 \div \frac{3}{5} = \frac{75}{1} \div \frac{3}{5} = \frac{\overset{25}{\cancel{75}}}{1} \cdot \frac{5}{\cancel{3}}$$
$$= \frac{25 \cdot 5}{1 \cdot 1}$$
$$= \frac{125}{1} = 125$$

125 maintenance trucks can be filled.

21. $5\frac{1}{4} \cdot 3\frac{3}{8} = \frac{21}{1} \cdot \frac{27}{8} = \frac{21 \cdot 27}{1 \cdot 8} = \frac{567}{8} = 17\frac{23}{32}$

22. $1\frac{5}{6} \cdot 4\frac{1}{3} = \frac{11}{6} \cdot \frac{13}{3} = \frac{11 \cdot 13}{6 \cdot 3} = \frac{143}{18} = 7\frac{17}{18}$

23. $4\frac{4}{5} \div 1\frac{1}{8} = \frac{24}{5} \div \frac{9}{8} = \frac{\overset{8}{\cancel{24}}}{5} \cdot \frac{8}{\cancel{9}}$
$$= \frac{8 \cdot 8}{5 \cdot 3}$$
$$= \frac{64}{15} = 4\frac{4}{15}$$

24. $\dfrac{9\frac{1}{2}}{3\frac{1}{3}} = 9\frac{1}{2} \div 3\frac{1}{3}$
$$= \frac{19}{2} \div \frac{10}{3}$$
$$= \frac{19}{2} \cdot \frac{3}{10}$$
$$= \frac{19 \cdot 3}{2 \cdot 10}$$
$$= \frac{57}{20} = 2\frac{17}{20}$$

25. Total grams can be found by multiplying the grams per day by the number of days.

$$1\frac{1}{4} \cdot 8\frac{1}{2} = \frac{5}{4} \cdot \frac{17}{2} = \frac{85}{8} = 10\frac{5}{8}$$

10 5/8 grams can be made.

CHAPTER 3 ADDING AND SUBTRACTING FRACTIONS

Section 3.1

3. $\dfrac{1}{10} + \dfrac{6}{10} = \dfrac{1+6}{10} = \dfrac{7}{10}$

7. $\dfrac{1}{9} + \dfrac{2}{9} = \dfrac{1+2}{9} = \dfrac{\overset{1}{\cancel{3}}}{\underset{3}{\cancel{9}}} = \dfrac{1}{3}$

11. $\dfrac{6}{20} + \dfrac{4}{20} + \dfrac{3}{20} = \dfrac{6+4+3}{20} = \dfrac{13}{20}$

15. $\dfrac{3}{8} + \dfrac{1}{8} + \dfrac{2}{8} = \dfrac{3+1+2}{8} = \dfrac{\overset{3}{\cancel{6}}}{\underset{4}{\cancel{8}}} = \dfrac{3}{4}$

19. $\dfrac{8}{11} - \dfrac{3}{11} = \dfrac{8-3}{11} = \dfrac{5}{11}$

23. $\dfrac{9}{10} - \dfrac{3}{10} = \dfrac{9-3}{10} = \dfrac{\overset{3}{\cancel{6}}}{\underset{5}{\cancel{10}}} = \dfrac{3}{5}$

27. $\dfrac{27}{40} - \dfrac{19}{40} = \dfrac{27-19}{40} = \dfrac{\overset{1}{\cancel{8}}}{\underset{5}{\cancel{40}}} = \dfrac{1}{5}$

31. $\dfrac{87}{144} - \dfrac{71}{144} = \dfrac{87-71}{144} = \dfrac{\overset{1}{\cancel{16}}}{\underset{9}{\cancel{144}}} = \dfrac{1}{9}$

35. Amount downhill + amount along creek = total walked.

$\dfrac{5}{12} + \dfrac{1}{12} = \dfrac{5+1}{12} = \dfrac{\overset{1}{\cancel{6}}}{\underset{2}{\cancel{12}}} = \dfrac{1}{2}$ mile

39. Amount planted in morning + amount planted in afternoon = total amount planted.

$\dfrac{5}{12} + \dfrac{11}{12} = \dfrac{16}{12}$

Total amount planted − amount destroyed = amount remaining.

$\dfrac{16}{12} - \dfrac{7}{12} = \dfrac{\overset{3}{\cancel{9}}}{\underset{4}{\cancel{12}}} = \dfrac{3}{4}$ acre

43.
```
        100
        / \
      10   10
      /\   /\
     2  5 2  5
```
$100 = 2 \cdot 2 \cdot 5 \cdot 5$

Section 3.2

3. Find the least common multiple of 12 and 15.

2	12	15
2	6	15
3	3	15
5	1	5
	1	1

Least common multiple
$= 2 \cdot 2 \cdot 3 \cdot 5 = 60$

7. Find the least common multiple of 25 and 40.

prime	2	5
25 =		5·5
40 =	2·2·2	5

The least common multiple is
$2 \cdot 2 \cdot 2 \cdot 5 \cdot 5 = 200$.

46 Chapter 3 Adding and Subtracting Fractions

11. Find the least common multiple of 12, 18, and 20.

2	12	18	20
2	6	9	10
3	3	9	5
3	1	3	5
5	1	1	5
	1	1	1

    Least common multiple
    $= 2 \cdot 2 \cdot 3 \cdot 3 \cdot 5 = 180$

15. Find the least common multiple of 18, 20, and 24.

prime	2	3	5
18 =	2	3·3	
20 =	2·2		5
24 =	2·2·2	3	

    The least common multiple is
    $2 \cdot 2 \cdot 2 \cdot 3 \cdot 3 \cdot 5 = 360$

19.
5	10	15	20	25
2	2	3	4	5
2	1	3	2	5
3	1	3	1	5
5	1	1	1	5
	1	1	1	1

    Least common multiple
    $= 2 \cdot 2 \cdot 3 \cdot 5 \cdot 5 = 300$

23.
10	10	30	50	60
3	1	3	5	6
2	1	1	5	2
5	1	1	5	1
	1	1	1	1

    Least common multiple
    $= 2 \cdot 3 \cdot 5 \cdot 10 = 300$

27. $\dfrac{3}{4} = \dfrac{}{36}$ $36 \div 4 = 9$

    $\dfrac{3 \cdot 9}{4 \cdot 9} = \dfrac{27}{36}$

31. $\dfrac{3}{8} = \dfrac{}{24}$ $24 \div 8 = 3$

    $\dfrac{3 \cdot 3}{8 \cdot 3} = \dfrac{9}{24}$

35. $\dfrac{7}{8} = \dfrac{}{56}$ $56 \div 8 = 7$

    $\dfrac{7 \cdot 7}{8 \cdot 7} = \dfrac{49}{56}$

39. $\dfrac{15}{19} = \dfrac{}{76}$ $76 \div 19 = 4$

    $\dfrac{15 \cdot 4}{19 \cdot 4} = \dfrac{60}{76}$

43. $\dfrac{8}{3} = \dfrac{}{51}$ $51 \div 3 = 17$

    $\dfrac{8 \cdot 17}{3 \cdot 17} = \dfrac{136}{51}$

47. $\dfrac{17}{800}$ and $\dfrac{23}{3600}$

    Find the least common multiple of 8 and 36 and multiply it by 100.

2	8	36 $= 2 \cdot 2 \cdot 2 \cdot 3 \cdot 3$
2	4	18 $= 72$
2	2	9
3	1	9 $72 \times 100 = 7200$
3	1	3
	1	1

51. $\dfrac{7}{5}$,

    $5 \overline{)7} \quad \begin{array}{l} 1 \text{ R2} \\ \underline{5} \\ 2, \end{array}$ Divide 7 by 5

    $1\dfrac{2}{5}$ ← Remainder
    $\dfrac{}{5}$ ← Divisor

3.3 Adding and Subtracting Unlike Fractions   47

55. $\dfrac{14}{11}$, $11\overline{)14}^{\ 1\ R3}$, $1\dfrac{3}{11}$
   $\phantom{11\overline{)14}}\underline{11}$
   $\phantom{11\overline{)1}}3$

Section 3.3

3. $\dfrac{9}{16} + \dfrac{3}{16} = \dfrac{9+3}{16} = \dfrac{\overset{3}{\cancel{12}}}{\underset{4}{\cancel{16}}} = \dfrac{3}{4}$

7. $\dfrac{9}{11} + \dfrac{1}{22}$ L.C.D. is 22

   $= \dfrac{9 \cdot 2}{11 \cdot 2} + \dfrac{1}{22}$

   $= \dfrac{18}{22} + \dfrac{1}{22}$

   $= \dfrac{18 + 1}{22} = \dfrac{19}{22}$

11. $\dfrac{1}{6} + \dfrac{5}{9}$ L.C.D. is 18

    $= \dfrac{1 \cdot 3}{6 \cdot 3} + \dfrac{5 \cdot 2}{9 \cdot 2}$

    $= \dfrac{3}{18} + \dfrac{10}{18}$

    $= \dfrac{3 + 10}{10} = \dfrac{13}{18}$

15. $\dfrac{3}{10} + \dfrac{2}{5} + \dfrac{3}{20}$ L.C.D. is 20

    $= \dfrac{3 \cdot 2}{10 \cdot 2} + \dfrac{2 \cdot 4}{5 \cdot 4} + \dfrac{3}{20}$

    $= \dfrac{6}{20} + \dfrac{8}{20} + \dfrac{3}{20}$

    $= \dfrac{6 + 8 + 3}{20} = \dfrac{17}{20}$

19. The least common denominator is 12.

    $\begin{array}{r}\dfrac{1}{4} = \dfrac{3}{12}\\+\dfrac{2}{3} = \dfrac{8}{12}\\\hline \dfrac{3+8}{12} = \dfrac{11}{12}\end{array}$

23. $\dfrac{5}{6} - \dfrac{1}{6} = \dfrac{5-1}{6} = \dfrac{\overset{2}{\cancel{4}}}{\underset{3}{\cancel{6}}} = \dfrac{2}{3}$

27. $\dfrac{5}{12} - \dfrac{1}{4}$ L.C.D. is 12

    $= \dfrac{5}{12} - \dfrac{1 \cdot 3}{4 \cdot 3}$

    $= \dfrac{5}{12} - \dfrac{3}{12}$

    $= \dfrac{5 - 3}{12} = \dfrac{\overset{1}{\cancel{2}}}{\underset{6}{\cancel{12}}} = \dfrac{1}{6}$

31. $\dfrac{8}{9} - \dfrac{7}{15}$ L.C.D. is 45

    $= \dfrac{8 \cdot 5}{9 \cdot 5} - \dfrac{7 \cdot 3}{15 \cdot 3} = \dfrac{40}{45} - \dfrac{21}{45}$

    $= \dfrac{40 - 21}{45} = \dfrac{19}{45}$

35. The least common denominator is 15.

    $\begin{array}{r}\dfrac{2}{3} = \dfrac{10}{15}\\-\dfrac{3}{5} = \dfrac{9}{15}\\\hline \dfrac{10-9}{15} = \dfrac{1}{15}\end{array}$

39. Acres the company owned − acres sold
    = acres left.

    $\dfrac{3}{4} - \dfrac{1}{6} = \dfrac{3 \cdot 3}{4 \cdot 3} - \dfrac{1 \cdot 2}{6 \cdot 2}$ L.C.D. is 12

    $= \dfrac{9}{12} - \dfrac{2}{12} = \dfrac{9-2}{12}$

    $= \dfrac{7}{12}$ acre left

43. $7\dfrac{1}{2} \cdot 3\dfrac{1}{3} = \dfrac{\overset{5}{\cancel{15}}}{\underset{1}{\cancel{2}}} \cdot \dfrac{\overset{5}{\cancel{10}}}{\underset{1}{\cancel{3}}}$

    $= \dfrac{5 \cdot 5}{1 \cdot 1} = \dfrac{25}{1} = 25$

48   Chapter 3   Adding and Subtracting Fractions

47. $1\frac{1}{2} \div 3\frac{3}{4} = \frac{3}{2} \div \frac{15}{4}$

    $= \frac{\cancel{3}^{1}}{\cancel{2}_{1}} \cdot \frac{\cancel{4}^{2}}{\cancel{15}_{5}} = \frac{1 \cdot 2}{1 \cdot 5} = \frac{2}{5}$

Section 3.4

3. $51\frac{1}{4} = 51\frac{1}{4}$
    $+\ 29\frac{1}{2} = 29\frac{2}{4}$
    $\overline{\phantom{+\ 29\frac{1}{2}}\quad 80\frac{1+2}{4} = 80\frac{3}{4}}$

    Add the whole numbers and add the fractions

7. $82\frac{3}{5}$
    $+\ 15\frac{4}{5}$
    $\overline{\phantom{+\ 15\frac{4}{5}}\ 97\frac{7}{5} = 97 + 1\frac{2}{5} = 98\frac{2}{5}}$

11. $268\frac{9}{10} = 268\frac{36}{40}$
    $+\ 35\frac{3}{8} = 35\frac{15}{40}$
    $\overline{\phantom{+\ 35\frac{3}{8}}\ 303\frac{51}{40} = 303 + 1\frac{11}{40}}$
    $\phantom{+\ 35\frac{3}{8}\ 303\frac{51}{40}} = 304\frac{11}{40}$

15. $28\frac{1}{4} = 28\frac{5}{20}$
    $23\frac{3}{5} = 23\frac{12}{20}$
    $+\ 19\frac{9}{10} = 19\frac{18}{20}$
    $\overline{\phantom{+\ 19\frac{9}{10}}\ 70\frac{35}{20} = 70 + 1\frac{15}{20}}$
    $\phantom{+\ 19\frac{9}{10}\ 70\frac{35}{20}} = 71\frac{15}{20} = 71\frac{3}{4}$

19. $11\frac{9}{20} = 11\frac{9}{20} = 10\frac{29}{20}$
    $-\ 4\frac{3}{5} = 4\frac{12}{20} = 4\frac{12}{20}$
    $\overline{\phantom{-\ 4\frac{3}{5} = 4\frac{12}{20} =\ }6\frac{17}{20}}$

    Subtract the whole numbers and subtract the fractions

23. $37\frac{1}{2} = 37\frac{4}{8} = 36\frac{12}{8}$
    $-\ 24\frac{5}{8} = 24\frac{5}{8} = 24\frac{5}{8}$
    $\overline{\phantom{-\ 24\frac{5}{8} = 24\frac{5}{8} =\ }12\frac{7}{8}}$

27. $26\frac{5}{18} = 26\frac{20}{72} = 25\frac{92}{72}$
    $-\ 12\frac{11}{24} = 12\frac{33}{72} = 12\frac{33}{72}$
    $\overline{\phantom{-\ 12\frac{11}{24} = 12\frac{33}{72} =\ }13\frac{59}{72}}$

31. $746\frac{3}{8} = 746\frac{9}{24} = 745\frac{33}{24}$
    $-\ 423\frac{11}{12} = 423\frac{22}{24} = 423\frac{22}{24}$
    $\overline{\phantom{-\ 423\frac{11}{12} = 423\frac{22}{24} =\ }322\frac{11}{24}}$

35. $115\phantom{\frac{15}{16}} = 114\frac{16}{16}$
    $-\ 62\frac{15}{16} = 62\frac{15}{16}$
    $\overline{\phantom{-\ 62\frac{15}{16} =\ }52\frac{1}{16}}$

39. Add the two lengths of wood.

    $12\frac{1}{2} = 12\frac{3}{6}$
    $+\ 7\frac{2}{3} = 7\frac{4}{6}$
    $\overline{\phantom{+\ 7\frac{2}{3}}\ 19\frac{7}{6} = 19 + 1\frac{1}{6} = 20\frac{1}{6}}$

    The total length of the wood is $20\frac{1}{6}$.

3.5 Order Relations and the Order of Operations   49

43. Add the amounts the driver gives out and subtract the answer from the amount the truck has.

$$1\tfrac{1}{2} = 1\tfrac{2}{4}$$
$$2\tfrac{3}{4} = 2\tfrac{3}{4}$$
$$+\ 3 = 3$$
$$6\tfrac{5}{4} = 6 + 1\tfrac{1}{4} = 7\tfrac{1}{4}$$

$$9\tfrac{5}{8} = 9\tfrac{5}{8}$$
$$-\ 7\tfrac{1}{4} = 7\tfrac{2}{8}$$
$$2\tfrac{3}{8} \text{ cubic yards left in truck}$$

47. Add the four amounts.

$$3\tfrac{1}{4} = 3\tfrac{6}{24}$$
$$2\tfrac{3}{8} = 2\tfrac{9}{24}$$
$$7\tfrac{1}{2} = 7\tfrac{12}{24}$$
$$+\ 1\tfrac{5}{6} = 1\tfrac{20}{24}$$
$$13\tfrac{47}{24} = 13 + 1\tfrac{23}{24}$$

$14\tfrac{23}{24}$ tons sold last month

51. $9^2 + 5 - 2$

    $= 81 + 5 - 2$ *Simplify exponents*
    $= 86 - 2$ *Add and subtract left to right*
    $= 84$ *Subtract last*

55. $3^2 \cdot (5 - 2)$

    $= 3^2 \cdot 3$ *Simplify inside parentheses*
    $= 9 \cdot 3$ *Simplify exponent*
    $= 27$ *Multiply last*

Section 3.5

For Exercises 3 and 7, see the number line graphs in the answer section of the textbook.

7. Note: $\tfrac{13}{6} = 2\tfrac{1}{6}$

11. $\tfrac{5}{8}$ — $\tfrac{11}{16}$

    $\tfrac{10}{16} < \tfrac{11}{16}$ *Common denominator is 16*

15. $\tfrac{7}{12}$ — $\tfrac{11}{18}$

    $\tfrac{21}{36} < \tfrac{22}{36}$ *Common denominator is 36*

19. $\tfrac{37}{50}$ — $\tfrac{13}{20}$

    $\tfrac{74}{100} > \tfrac{65}{100}$ *Common denominator is 100*

23. $\left(\tfrac{7}{15}\right)^2 = \tfrac{7}{15} \cdot \tfrac{7}{15} = \tfrac{49}{225}$

27. $\left(\tfrac{2}{9}\right)^3 = \tfrac{2}{9} \cdot \tfrac{2}{9} \cdot \tfrac{2}{9} = \tfrac{8}{729}$

31. $\left(\tfrac{1}{2}\right)^5 = \tfrac{1}{2} \cdot \tfrac{1}{2} \cdot \tfrac{1}{2} \cdot \tfrac{1}{2} \cdot \tfrac{1}{2} = \tfrac{1}{32}$

35. $8 \cdot 3^2 - \tfrac{10}{2}$

    $= 8 \cdot 9 - \tfrac{10}{2}$ *Evaluate exponents*
    $= 72 - \tfrac{10}{2}$ *Multiply*
    $= 72 - 5$ *Divide next*
    $= 67$ *Subtract*

Chapter 3 Adding and Subtracting Fractions

39. $\left(\frac{3}{4}\right)^2 \cdot \left(\frac{1}{3}\right)$

    $= \frac{3}{4} \cdot \frac{3}{4} \cdot \frac{1}{3}$

    $= \frac{\cancel{9}^{3}}{16} \cdot \frac{1}{\cancel{3}_{1}}$

    $= \frac{3}{16}$

43. $6 \cdot \left(\frac{2}{3}\right)^2 \cdot \left(\frac{1}{2}\right)^3$

    $= \frac{\cancel{6}}{1} \cdot \frac{4}{\cancel{9}} \cdot \frac{1}{8}$ (with 2 over 6 and 3 under 9)

    $= \frac{\cancel{8}^{1}}{3} \cdot \frac{1}{\cancel{8}_{1}}$

    $= \frac{1}{3}$

47. $\frac{1}{2} + \left(\frac{1}{2}\right)^2 - \frac{3}{8}$

    $= \frac{1}{2} + \frac{1}{4} - \frac{3}{8}$ *Evaluate exponents*

    $= \frac{4}{8} + \frac{2}{8} - \frac{3}{8}$ *Find common denominator*

    $= \frac{6}{8} - \frac{3}{8}$

    $= \frac{3}{8}$

51. $\frac{9}{8} \div \left(\frac{2}{3} + \frac{1}{12}\right)$

    $= \frac{9}{8} \div \left(\frac{8}{12} + \frac{1}{12}\right)$

    $= \frac{9}{8} \div \frac{9}{12}$

    $= \frac{\cancel{9}^{1}}{\cancel{8}_{2}} \cdot \frac{\cancel{12}^{3}}{\cancel{9}_{1}}$

    $= \frac{3}{2} = 1\frac{1}{2}$

55. $\frac{3}{8} \cdot \left(\frac{1}{4} + \frac{1}{2}\right) \cdot \frac{32}{3}$

    $= \frac{3}{8} \cdot \left(\frac{3}{4}\right) \cdot \frac{32}{3}$

    $= 3$

59. $\left(\frac{3}{5}\right)^2 \cdot \left(\frac{1}{3} + \frac{2}{9}\right) - \frac{1}{2} \cdot \frac{1}{5}$

    $= \left(\frac{9}{25}\right) \cdot \left(\frac{5}{9}\right) - \frac{1}{10}$ *Evaluate exponents*

    $= \frac{1}{5} - \frac{1}{10}$ *Multiply*

    $= \frac{1}{10}$ *Subtract*

63. four million, seventy-one thousand, two hundred eighty

    (Do not use "and eighty".)

Chapter 3 Review Exercises

1. $\frac{1}{2} + \frac{1}{2} = \frac{1+1}{2} = \frac{2}{2} = 1$

2. $\frac{3}{8} + \frac{4}{8} = \frac{3+4}{8} = \frac{7}{8}$

3. $\frac{1}{12} + \frac{2}{12} + \frac{1}{12} = \frac{1+2+1}{12}$

    $= \frac{4}{12} = \frac{1}{3}$

4. $\frac{8}{14} - \frac{3}{14} = \frac{8-3}{14} = \frac{5}{14}$

5. $\frac{3}{10} - \frac{1}{10} = \frac{3-1}{10} = \frac{2}{10} = \frac{1}{5}$

6. $\frac{8}{32} - \frac{4}{32} = \frac{8-4}{32} = \frac{4}{32} = \frac{1}{8}$

Chapter 3 Review Exercises 51

7. $\dfrac{36}{62} - \dfrac{10}{62} = \dfrac{36-10}{62} = \dfrac{26}{62} = \dfrac{13}{31}$

8. $\dfrac{208}{360} - \dfrac{170}{360} = \dfrac{208-170}{360}$
   $= \dfrac{38}{360} = \dfrac{19}{180}$

9. To find what portion John traveled in two days, add:

   $\dfrac{3}{8} + \dfrac{2}{8} = \dfrac{3+2}{8} = \dfrac{5}{8}.$

   He completed $\dfrac{5}{8}$ of his travel.

10. To find out how much less reading Diane did in the afternoon, subtract:

    $\dfrac{3}{7} - \dfrac{2}{7} = \dfrac{3-2}{7} = \dfrac{1}{7}.$

    Diane read $\dfrac{1}{7}$ less in the afternoon.

11.
    ```
    2 | 10   8
    2 |  5   4
    2 |  5   2
    5 |  5   1
    |  1   1
    ```
    L.C.M. = 2 · 2 · 2 · 5 = 40

12.
    ```
    2 | 5   12
    2 | 5    6
    3 | 5    3
    5 | 5    1
    |  1    1
    ```
    L.C.M. = 2 · 2 · 3 · 5 = 60

13.
    ```
    2 | 10   12   20
    2 |  5    6   10
    3 |  5    3    5
    5 |  5    1    5
    |  1    1    1
    ```
    L.C.M. = 2 · 2 · 3 · 5 = 60

14.
    ```
    2 | 9   20   15
    2 | 9   10   15
    3 | 9    5   15
    3 | 3    5    5
    5 | 1    5    5
    |  1    1    1
    ```
    L.C.M. = 2 · 2 · 3 · 3 · 5 = 180

15.
    ```
    2 | 6    8    5   15
    2 | 3    4    5   15
    2 | 3    2    5   15
    3 | 3    1    5   15
    5 | 1    1    5    5
    |  1    1    1    1
    ```
    L.C.M. = 2 · 2 · 2 · 3 · 5 = 120

16.
    ```
    2 | 25   16   5   18
    2 | 25    8   5    9
    2 | 25    4   5    9
    2 | 25    2   5    9
    3 | 25    1   5    9
    3 | 25    1   5    3
    5 | 25    1   5    1
    5 |  5    1   1    1
    |   1    1   1    1
    ```
    L.C.M. = 2 · 2 · 2 · 2 · 3 · 3 · 5 · 5
         = 3600

17. $\dfrac{2}{5} = \dfrac{}{25}$ $25 \div 5 = 5$

    $\dfrac{2 \cdot 5}{2 \cdot 5} = \dfrac{10}{25}$

18. $\dfrac{7}{12} = \dfrac{}{48}$ $48 \div 12 = 4$

    $\dfrac{7 \cdot 4}{12 \cdot 4} = \dfrac{28}{48}$

19. $\dfrac{5}{6} = \dfrac{}{102}$ $102 \div 6 = 17$

    $\dfrac{5 \cdot 17}{6 \cdot 17} = \dfrac{85}{102}$

Chapter 3   Adding and Subtracting Fractions

20. $\dfrac{5}{9} = \dfrac{}{81}$ $\quad 81 \div 9 = 9$

    $\dfrac{5 \cdot 9}{9 \cdot 9} = \dfrac{45}{81}$

21. $\dfrac{7}{16} = \dfrac{}{144}$ $\quad 144 \div 16 = 9$

    $\dfrac{7 \cdot 9}{16 \cdot 9} = \dfrac{63}{144}$

22. $\dfrac{3}{22} = \dfrac{}{88}$ $\quad 88 \div 22 = 4$

    $\dfrac{3 \cdot 4}{22 \cdot 4} = \dfrac{12}{88}$

23. $\dfrac{1}{7} + \dfrac{4}{7} = \dfrac{1+4}{7} = \dfrac{5}{7}$

24. $\dfrac{1}{5} + \dfrac{4}{15}$ L.C.M is 15

    $\dfrac{1 \cdot 3}{5 \cdot 3} + \dfrac{4}{15} = \dfrac{3}{15} + \dfrac{4}{15}$

    $= \dfrac{3+4}{15} = \dfrac{7}{15}$

25. $\dfrac{3}{10} + \dfrac{1}{2} + \dfrac{1}{5}$ L.C.M. is 10

    $\dfrac{3}{10} + \dfrac{1 \cdot 5}{2 \cdot 5} + \dfrac{1 \cdot 2}{5 \cdot 2} = \dfrac{3}{10} + \dfrac{5}{10} + \dfrac{2}{10}$

    $= \dfrac{3+5+2}{10}$

    $= \dfrac{10}{10} = 1$

26. $\dfrac{1}{2} + \dfrac{3}{8} + \dfrac{1}{16}$ L.C.M is 16

    $\dfrac{1 \cdot 8}{2 \cdot 8} + \dfrac{3 \cdot 2}{8 \cdot 2} + \dfrac{1}{16} = \dfrac{8}{16} + \dfrac{6}{16} + \dfrac{1}{16}$

    $= \dfrac{8+6+1}{16}$

    $= \dfrac{15}{16}$

27. $\dfrac{1}{4} = \dfrac{3}{12}$ L.C.M is 12

    $+\dfrac{2}{3} = \dfrac{8}{12}$

    $\dfrac{3+8}{12} = \dfrac{11}{12}$

28. $\dfrac{5}{9} = \dfrac{20}{36}$

    $+\dfrac{1}{12} = \dfrac{3}{36}$

    $\dfrac{20+3}{36} = \dfrac{23}{36}$

29. $\dfrac{9}{16} = \dfrac{27}{48}$

    $+\dfrac{1}{12} = \dfrac{4}{48}$

    $\dfrac{27+4}{48} = \dfrac{31}{48}$

30. $\dfrac{4}{9} - \dfrac{2}{9} = \dfrac{4-2}{9} = \dfrac{2}{9}$

31. $\dfrac{7}{8} - \dfrac{7}{16}$

    $\dfrac{7 \cdot 2}{8 \cdot 2} - \dfrac{7}{16} = \dfrac{14}{16} - \dfrac{7}{16}$

    $= \dfrac{14-7}{16} = \dfrac{7}{16}$

32. $\dfrac{7}{16} = \dfrac{7}{16}$

    $-\dfrac{1}{4} = \dfrac{4}{16}$

    $\dfrac{7-4}{16} = \dfrac{3}{16}$

33. $\dfrac{5}{8} = \dfrac{15}{24}$

    $-\dfrac{1}{3} = \dfrac{8}{24}$

    $\dfrac{15-8}{24} = \dfrac{7}{24}$

34. $\dfrac{8}{15} = \dfrac{16}{30}$

    $-\dfrac{3}{10} = \dfrac{9}{30}$

    $\dfrac{16-9}{30} = \dfrac{7}{30}$

Chapter 3 Review Exercises   53

35. To find total cubic yards of gravel, add:

$$\frac{1}{4} + \frac{1}{3} + \frac{3}{8} = \frac{1 \cdot 6}{4 \cdot 6} + \frac{1 \cdot 8}{3 \cdot 8} + \frac{3 \cdot 3}{8 \cdot 3}$$

$$= \frac{6}{24} + \frac{8}{24} + \frac{9}{24}$$

$$= \frac{6 + 8 + 9}{24} = \frac{23}{24}.$$

The truck contains $\frac{23}{24}$ cubic yard of gravel.

36. To find the total portion he has installed, add:

$$\frac{3}{8} + \frac{1}{3} + \frac{1}{5} = \frac{3 \cdot 15}{8 \cdot 15} + \frac{1 \cdot 40}{3 \cdot 40} + \frac{1 \cdot 24}{5 \cdot 24}$$

$$= \frac{45}{120} + \frac{40}{120} + \frac{24}{120}$$

$$= \frac{45 + 40 + 24}{120}$$

$$= \frac{109}{120}.$$

Dick has installed $\frac{109}{120}$ of the thermostats.

37. $\quad 8\frac{1}{4}$
   $+ 9\frac{3}{4}$
   $\overline{\quad 17\frac{4}{4} = 17 + 1 = 18}$

38. $\quad 25\frac{3}{4} = 25\frac{6}{8}$
   $+ 16\frac{3}{8} = 16\frac{3}{8}$
   $\overline{\qquad 41\frac{9}{8} = 41 + \frac{1}{8} = 42\frac{1}{8}}$

39. $\quad 78\frac{3}{7}$
   $+ 17\frac{6}{7}$
   $\overline{\quad 95\frac{9}{7} = 95 + 1\frac{2}{7} = 96\frac{2}{7}}$

40. $\quad 12\frac{3}{5} \ = \ 12\frac{48}{80}$
   $\qquad 8\frac{5}{8} \ = \ 8\frac{50}{80}$
   $+ 10\frac{5}{16} \ = \ 10\frac{25}{80}$
   $\overline{\qquad 30\frac{123}{80} = 30 + 1\frac{43}{80} = 31\frac{43}{80}}$

41. $\quad 6\frac{2}{3} \ = \ 6\frac{4}{6}$
   $- 1\frac{1}{2} \ = \ 1\frac{3}{6}$
   $\overline{\qquad\qquad 5\frac{1}{6}}$

42. $\quad 17\frac{1}{2} \ = \ 17\frac{3}{6}$
   $- 11\frac{1}{3} \ = \ 11\frac{2}{6}$
   $\overline{\qquad\qquad 6\frac{1}{6}}$

43. $\quad 73 \ = \ 72\frac{3}{3}$
   $- 55\frac{2}{3} \ = \ 55\frac{2}{3}$
   $\overline{\qquad\qquad 17\frac{1}{3}}$

44. $\quad 238\frac{1}{8} \ = \ 237\frac{9}{8}$
   $- 152\frac{3}{8} \ = \ 152\frac{3}{8}$
   $\overline{\qquad\qquad 85\frac{6}{8} = 85\frac{3}{4}}$

45. To find the gallons remaining, subtract:

$$14\tfrac{1}{3} = 14\tfrac{2}{6} = 13\tfrac{8}{6}$$
$$-\ 5\tfrac{1}{2} = 5\tfrac{3}{6} = 5\tfrac{3}{6}$$
$$\overline{\ 8\tfrac{5}{6}}$$

The lab has $8\tfrac{5}{6}$ gallons of distilled water remaining.

46. To find the total amount, add:

$$6\tfrac{4}{5} = 6\tfrac{12}{15}$$
$$+\ 9\tfrac{2}{3} = 9\tfrac{10}{15}$$
$$\overline{\ 15\tfrac{22}{15} = 15 + 1\tfrac{1}{15} = 16\tfrac{7}{15}.}$$

The Scouts collected $16\tfrac{7}{15}$ tons of newspaper.

47. To find their total weight, add:

$$5\tfrac{3}{4} = 5\tfrac{18}{24}$$
$$4\tfrac{7}{8} = 4\tfrac{21}{24}$$
$$+\ 5\tfrac{1}{3} = 5\tfrac{8}{24}$$
$$\overline{\ 14\tfrac{47}{24} = 14 + 1\tfrac{23}{24} = 15\tfrac{23}{24}.}$$

The triplets' total weight is $15\tfrac{23}{24}$ pounds.

48. To find the total amount she has, add the two parcels.

$$1\tfrac{11}{16} = 1\tfrac{11}{16}$$
$$+\ 2\tfrac{3}{4} = 2\tfrac{12}{16}$$
$$\overline{\ 3\tfrac{23}{16} = 3 + 1\tfrac{7}{16} = 4\tfrac{7}{16}}$$

To find how much more she needs, subtract the total amount she has from the total amount she needs.

$$8\tfrac{1}{2} = 8\tfrac{8}{16}$$
$$-\ 4\tfrac{7}{16} = 4\tfrac{7}{16}$$
$$\overline{\ 4\tfrac{1}{16}}$$

She needs $4\tfrac{1}{16}$ more acres of land.

49–52. For Exercises 49–52, see the number line graphs in the answer section of the textbook.

53. $\tfrac{1}{2}$ — $\tfrac{5}{8}$

$\tfrac{4}{8} < \tfrac{5}{8}$ Find common denominator

54. $\tfrac{2}{3}$ — $\tfrac{5}{6}$

$\tfrac{4}{6} < \tfrac{5}{6}$ Find common denominator

55. $\tfrac{5}{6}$ — $\tfrac{7}{9}$

$\tfrac{15}{18} > \tfrac{14}{18}$ Find common denominator

56. $\tfrac{7}{10}$ — $\tfrac{8}{15}$

$\tfrac{21}{30} > \tfrac{16}{30}$ Find common denominator

57. $\tfrac{5}{12}$ — $\tfrac{8}{18}$

$\tfrac{15}{36} < \tfrac{16}{36}$ Find common denominator

58. $\dfrac{19}{25}$ —— $\dfrac{23}{30}$

$\dfrac{114}{150} < \dfrac{115}{150}$ Find common denominator

59. $\dfrac{19}{36}$ —— $\dfrac{29}{54}$

$\dfrac{114}{216} < \dfrac{116}{216}$ Find common denominator

60. $\dfrac{19}{132}$ —— $\dfrac{7}{55}$

$\dfrac{95}{660} > \dfrac{64}{660}$ Find common denominator

61. $\left(\dfrac{1}{3}\right)^2 = \dfrac{1}{3} \cdot \dfrac{1}{3} = \dfrac{1}{9}$

62. $\left(\dfrac{3}{8}\right)^2 = \dfrac{3}{8} \cdot \dfrac{3}{8} = \dfrac{9}{64}$

63. $\left(\dfrac{3}{5}\right)^3 = \dfrac{3}{5} \cdot \dfrac{3}{5} \cdot \dfrac{3}{5} = \dfrac{27}{125}$

64. $\left(\dfrac{3}{8}\right)^4 = \dfrac{3}{8} \cdot \dfrac{3}{8} \cdot \dfrac{3}{8} \cdot \dfrac{3}{8} = \dfrac{81}{4096}$

65. $\left(\dfrac{1}{3}\right)^2 \cdot 6 = \dfrac{1}{9} \cdot \dfrac{6}{1} = \dfrac{6}{9} = \dfrac{2}{3}$

66. $\left(\dfrac{3}{8}\right)^2 \cdot 16 = \dfrac{9}{\cancel{64}} \cdot \dfrac{\cancel{16}^{\,1}}{1} = \dfrac{9}{4} = 2\dfrac{1}{4}$
4

67. $\left(\dfrac{1}{5}\right)^2 \cdot \left(\dfrac{10}{7}\right)^2 = \dfrac{1}{\cancel{25}} \cdot \dfrac{\cancel{100}^{\,4}}{49} = \dfrac{4}{49}$
1

68. $\dfrac{3}{5} \div \left(\dfrac{1}{10} + \dfrac{1}{5}\right) = \dfrac{3}{5} \div \left(\dfrac{1}{10} + \dfrac{2}{10}\right)$

$ = \dfrac{3}{5} \div \dfrac{3}{10}$

$ = \dfrac{3}{5} \cdot \dfrac{10}{3}$

$ = \dfrac{30}{15} = 2$

69. $\left(\dfrac{1}{2}\right)^2 \cdot \left(\dfrac{1}{4} + \dfrac{1}{2}\right) = \dfrac{1}{4} \cdot \left(\dfrac{1}{4} + \dfrac{2}{4}\right)$

$ = \dfrac{1}{4} \cdot \dfrac{3}{4}$

$ = \dfrac{3}{16}$

70. $\left(\dfrac{1}{2}\right)^3 + \left(\dfrac{1}{4} + \dfrac{1}{12}\right) \div \dfrac{1}{9} \cdot \dfrac{3}{4}$

$= \dfrac{1}{8} + \left(\dfrac{3}{12} + \dfrac{1}{12}\right) \div \dfrac{1}{9} \cdot \dfrac{3}{4}$

$= \dfrac{1}{8} + \dfrac{4}{12} \div \dfrac{1}{9} \cdot \dfrac{3}{4}$

$= \dfrac{1}{8} + \dfrac{4}{12} \cdot \dfrac{9}{1} \cdot \dfrac{3}{4}$

$= \dfrac{1}{8} + \dfrac{9}{4}$

$= \dfrac{1}{8} + \dfrac{18}{8} = \dfrac{19}{8}$

$= 2\dfrac{3}{8}$

71. $\dfrac{9}{15} + \dfrac{4}{15} + \dfrac{9+4}{15} = \dfrac{13}{15}$

72. $\dfrac{3}{4} - \dfrac{1}{8}$ L.C.D is 8

$\dfrac{3 \cdot 2}{4 \cdot 2} - \dfrac{1}{8} = \dfrac{6}{8} - \dfrac{1}{8}$

$ = \dfrac{6-1}{8} = \dfrac{5}{8}$

73. $\dfrac{75}{86} - \dfrac{8}{86} = \dfrac{75-8}{86} = \dfrac{67}{86}$

74. $\frac{1}{4} + \frac{1}{8} + \frac{5}{16}$

$\frac{1 \cdot 4}{4 \cdot 4} + \frac{1 \cdot 2}{8 \cdot 2} + \frac{5}{16} = \frac{4}{16} + \frac{2}{16} + \frac{5}{16}$

$\qquad = \frac{4 + 2 + 5}{16} = \frac{11}{16}$

75. $9\frac{1}{3} = 9\frac{8}{24} = 8\frac{32}{24}$
$-5\frac{5}{8} = 5\frac{15}{24} = 5\frac{15}{24}$
$\qquad\qquad\qquad\qquad 3\frac{17}{24}$

76. $8\frac{3}{4} = 8\frac{3}{4}$
$+15\frac{1}{2} = 15\frac{2}{4}$
$\qquad\qquad 23\frac{5}{4} = 23 + 1\frac{1}{4} = 24\frac{1}{4}$

77. $\frac{7}{10} = \frac{28}{40}$
$-\frac{3}{8} = \frac{15}{40}$
$\qquad\quad \frac{28 - 15}{40} = \frac{13}{40}$

78. $12\frac{3}{5} = 12\frac{48}{80}$
$8\frac{5}{8} = 8\frac{50}{80}$
$+10\frac{5}{16} = 10\frac{25}{80}$
$\qquad\qquad 30\frac{123}{80} = 30 + 1\frac{43}{80} = 31\frac{43}{80}$

79. $936\frac{1}{2} = 936\frac{2}{4} = 935\frac{6}{4}$
$-618\frac{3}{4} = 618\frac{3}{4} = 618\frac{3}{4}$
$\qquad\qquad\qquad\qquad\qquad 317\frac{3}{4}$

80. $\frac{7}{22} + \frac{3}{22} + \frac{6}{22} = \frac{7 + 3 + 6}{22}$
$\qquad\qquad\qquad\qquad = \frac{16}{22} = \frac{8}{11}$

81. $\left(\frac{1}{4}\right)^2 \cdot \left(\frac{2}{5}\right)^3 = \frac{1}{\cancel{16}} \cdot \frac{\cancel{8}^1}{125} = \frac{1}{250}$
$\qquad\qquad\qquad\qquad\quad 2$

82. $\frac{1}{4} \div \left(\frac{1}{3} + \frac{1}{6}\right) = \frac{1}{4} \div \left(\frac{2}{6} + \frac{1}{6}\right)$
$\qquad\qquad\qquad = \frac{1}{4} \div \frac{3}{6}$
$\qquad\qquad\qquad = \frac{1}{4} \cdot \frac{6}{3} = \frac{6}{12}$
$\qquad\qquad\qquad = \frac{1}{2}$

83. $\left(\frac{2}{3}\right)^2 \cdot \left(\frac{1}{3} + \frac{1}{6}\right) = \frac{4}{9} \cdot \left(\frac{2}{6} + \frac{1}{6}\right)$
$\qquad\qquad\qquad = \frac{4}{9} \cdot \frac{3}{6} = \frac{12}{54}$
$\qquad\qquad\qquad = \frac{2}{9}$

84. $\frac{11}{9} \underline{\quad} \frac{11}{6}$

$\frac{22}{18} < \frac{33}{18}$ Find common denominator

85. $\frac{10}{11} \underline{\quad} \frac{32}{33}$

$\frac{30}{33} < \frac{32}{33}$ Find common denominator

86. $\frac{19}{40} \underline{\quad} \frac{29}{60}$

$\frac{57}{120} < \frac{58}{120}$ Find common denominator

87. $\frac{17}{12} \underline{\quad} \frac{25}{54}$

$\frac{153}{108} > \frac{50}{108}$ Find common denominator

Chapter 3 Test 57

88.
```
 2 | 12   22
 2 |  6   11
 3 |  3   11
11 |  1   11
        1    1
```
The least common multiple is
$2 \cdot 2 \cdot 3 \cdot 11 = 132$.

89.
```
2 | 2   16   36   42
2 | 1    8   18   21
2 | 1    4    9   21
2 | 1    2    9   21
3 | 1    1    9   21
3 | 1    1    3    7
7 | 1    1    1    7
    1    1    1    1
```
The least common multiple is
$2 \cdot 2 \cdot 2 \cdot 2 \cdot 3 \cdot 3 \cdot 7 = 1008$.

90. $\dfrac{3}{7} = \dfrac{}{560}$ $560 \div 7 = 80$

$\dfrac{3 \cdot 80}{7 \cdot 80} = \dfrac{240}{560}$

91. $\dfrac{9}{12} = \dfrac{}{144}$ $144 \div 12 = 12$

$\dfrac{9 \cdot 12}{12 \cdot 12} = \dfrac{108}{144}$

92. To find the total length, add.

$$\begin{array}{r} 21\frac{5}{16} = 21\frac{5}{16} \\ + 7\frac{3}{8} = 7\frac{6}{16} \\ \hline 28\frac{11}{16} \end{array}$$

The plumber needs $28\frac{11}{16}$ inches of pipe.

93. To find the total number of positions already accounted for, add.

$$\begin{array}{r} 1\frac{5}{8} = 1\frac{15}{24} \\ + 4\frac{5}{6} = 4\frac{20}{24} \\ \hline 5\frac{35}{24} = 5 + 1\frac{11}{24} = 6\frac{11}{24} \end{array}$$

To find the number of positions that remain, subtract the positions already accounted for from the total number of positions to be filled.

$$\begin{array}{r} 10 = 9\frac{24}{24} \\ - 6\frac{11}{24} = 6\frac{11}{24} \\ \hline 3\frac{13}{24} \end{array}$$

$3\frac{13}{24}$ positions remain to be filled.

Chapter 3 Test

1. $\dfrac{5}{8} + \dfrac{1}{8} = \dfrac{5+1}{8}$

    $= \dfrac{6}{8} = \dfrac{3}{4}$

2. $\dfrac{3}{10} + \dfrac{5}{10} = \dfrac{3+5}{10}$

    $= \dfrac{8}{10} = \dfrac{4}{5}$

3. $\dfrac{4}{5} - \dfrac{3}{5} = \dfrac{4-3}{5}$

    $= \dfrac{1}{5}$

4. $\dfrac{9}{15} - \dfrac{6}{15} = \dfrac{9-6}{15} = \dfrac{3}{15}$

    $= \dfrac{1}{5}$

5.
```
2 | 4   8   2   16
2 | 2   4   1   8
2 | 1   2   1   4
2 | 1   1   1   2
    1   1   1   1
```

Least common multiple = $2 \cdot 2 \cdot 2 \cdot 2$
$= 16$

6.
```
3 | 7   15   3   5
5 | 7    5   1   5
7 | 7    1   1   1
    1    1   1   1
```

Least common multiple = $3 \cdot 5 \cdot 7 = 105$

7.
```
3 | 3   5   7   9
3 | 1   5   7   3
5 | 1   5   7   1
7 | 1   1   7   1
    1   1   1   1
```

Least common multiple = $3 \cdot 3 \cdot 5 \cdot 7$
$= 315$

8. $\dfrac{7}{16} + \dfrac{2}{3} = \dfrac{7 \cdot 3}{16 \cdot 3} + \dfrac{2 \cdot 16}{3 \cdot 16}$

   $= \dfrac{21}{48} + \dfrac{32}{48} = \dfrac{53}{48} = 1\dfrac{5}{48}$

9. $\dfrac{4}{5} + \dfrac{3}{7} = \dfrac{4 \cdot 7}{5 \cdot 7} + \dfrac{3 \cdot 5}{7 \cdot 5}$

   $= \dfrac{28}{35} + \dfrac{15}{35} = \dfrac{43}{35} = 1\dfrac{8}{35}$

10. $\dfrac{5}{9} - \dfrac{1}{6} = \dfrac{5 \cdot 2}{9 \cdot 2} - \dfrac{1 \cdot 3}{6 \cdot 3}$

    $= \dfrac{10}{18} - \dfrac{3}{18} = \dfrac{7}{18}$

11. $\dfrac{7}{8} - \dfrac{6}{7} = \dfrac{7 \cdot 7}{8 \cdot 7} - \dfrac{6 \cdot 8}{7 \cdot 8}$

    $= \dfrac{49}{56} - \dfrac{48}{56} = \dfrac{1}{56}$

12. $ 1\dfrac{1}{2} = 1\dfrac{2}{4}$
    $+ 3\dfrac{1}{4} = 3\dfrac{1}{4}$
    $\phantom{+ 3\dfrac{1}{4} =\ } 4\dfrac{3}{4}$

13. $ 5\dfrac{7}{8} = 5\dfrac{7}{8}$
    $+ 2\dfrac{3}{4} = 2\dfrac{6}{8}$
    $\phantom{+ 2\dfrac{3}{4} =\ } 7\dfrac{13}{8} = 7 + 1\dfrac{5}{8} = 8\dfrac{5}{8}$

14. $ 9\dfrac{3}{4} = 9\dfrac{15}{20}$
    $- 4\dfrac{3}{10} = 4\dfrac{6}{20}$
    $\phantom{- 4\dfrac{3}{10} =\ } 5\dfrac{9}{20}$

15. $ 7\dfrac{2}{3} = 7\dfrac{8}{12} = 6\dfrac{20}{12}$
    $- 4\dfrac{11}{12} = 4\dfrac{11}{12} = 4\dfrac{11}{12}$
    $\phantom{- 4\dfrac{11}{12} = 4\dfrac{11}{12} =\ } 2\dfrac{9}{12} = 2\dfrac{3}{4}$

16. $ 18\dfrac{3}{4} = 18\dfrac{45}{60}$
    $\ \ 9\dfrac{2}{5} = 9\dfrac{24}{60}$
    $+ 12\dfrac{1}{3} = 12\dfrac{20}{60}$
    $\phantom{+ 12\dfrac{1}{3} =\ } 39\dfrac{89}{60} = 39 + 1\dfrac{29}{60} = 40\dfrac{29}{60}$

17. $ 276\dfrac{1}{4} = 276\dfrac{2}{8} = 275\dfrac{10}{8}$
    $- 127\dfrac{5}{8} = 127\dfrac{5}{8} = 127\dfrac{5}{8}$
    $\phantom{- 127\dfrac{5}{8} = 127\dfrac{5}{8} =\ } 148\dfrac{5}{8}$

Chapter 3 Test   59

18. To find the total, add:

$$3\tfrac{1}{4} = 3\tfrac{3}{12}$$
$$4\tfrac{1}{6} = 4\tfrac{2}{12}$$
$$2\tfrac{1}{3} = 2\tfrac{4}{12}$$
$$3\tfrac{5}{6} = 3\tfrac{10}{12}$$
$$+\,4\tfrac{2}{3} = 4\tfrac{8}{12}$$
$$\overline{16\tfrac{27}{12}} = 16 + 2\tfrac{3}{12}$$
$$= 18\tfrac{3}{12} = 18\tfrac{1}{4}.$$

He studied for $18\tfrac{1}{4}$ hours.

19. To find how much more is needed, subtract.

$$12\tfrac{1}{4} = 12\tfrac{3}{12} = 11\tfrac{15}{12}$$
$$-\,6\tfrac{2}{3} = 6\tfrac{8}{12} = 6\tfrac{8}{12}$$
$$\overline{5\tfrac{7}{12}}$$

The club needs $5\tfrac{7}{12}$ tons to fill their dumpster.

20. $\tfrac{3}{5} \;\text{—}\; \tfrac{13}{20}$

$\tfrac{12}{20} < \tfrac{13}{20}$ *Find common denominator*

21. $\tfrac{11}{18} \;\text{—}\; \tfrac{17}{24}$

$\tfrac{44}{72} < \tfrac{51}{72}$ *Find common denominator*

22. $\left(\tfrac{1}{2}\right)^2 \cdot 2$

$= \tfrac{1}{4} \cdot 2$

$= \tfrac{1}{\cancel{4}} \cdot \tfrac{\cancel{2}^{\,1}}{1}$
${}_{2}$

$= \tfrac{1}{2}$

23. $\left(\tfrac{5}{8}\right)^2 \cdot \left(\tfrac{2}{3}\right)^2$

$= \tfrac{25}{\cancel{64}} \cdot \tfrac{\cancel{4}^{\,1}}{9}$
${}_{16}$

$= \tfrac{25}{144}$

24. $\left(\tfrac{5}{6} - \tfrac{5}{12}\right) \cdot 3$

$= \left(\tfrac{10}{12} - \tfrac{5}{12}\right) \cdot 3$

$= \tfrac{5}{12} \cdot 3$

$= \tfrac{5}{\cancel{12}} \cdot \tfrac{\cancel{3}^{\,1}}{1}$
${}_{4}$

$= \tfrac{5}{4} = 1\tfrac{1}{4}$

25. $\tfrac{2}{3} + \tfrac{5}{8} \cdot \tfrac{4}{3} = \tfrac{2}{3} + \tfrac{5}{6}$ *Multiply*

$= \tfrac{4}{6} + \tfrac{5}{6} = \tfrac{9}{6} = \tfrac{3}{2} = 1\tfrac{1}{2}$

Chapter 3 Adding and Subtracting Fractions

Cumulative Review Exercises
Chapters 1–3

1. 946
   └── 6 in the ones place
   └── 9 in the hundreds place

2. 8,354,917
   └── 4 in the thousands place
   └── 8 in the millions place

3.
   ```
     7
     6
     4
   + 9
    26
   ```

4.
   ```
     15
     28
     38
   + 42
    123
   ```
   └── Write 3, carry 2
   └── Write 12

5.
   ```
    51,506
     9 834
       279
   + 15,702
    77,321
   ```
   └── Write 1, carry 2
   └── Write 2, carry 1
   └── Write 3, carry 2
   └── Write 7, carry 1

6.
   ```
     375,899
     521,742
   + 357,968
   1,255,609
   ```
   └── Write 9, carry 1
   └── Write 0, carry 2
   └── Write 6, carry 2
   └── Write 5, carry 1
   └── Write 5, carry 1
   └── Write 12

7.
   ```
         11
      6̶ 12
      7̶2̶ 2̶
    - 54 6
      17 6
   ```

8.
   ```
         11
      2̶ 14
      3̶2̶4̶ 6
    - 298 3
       26 3
   ```

9.
   ```
       11 14 10
      1̶2̶ 8̶ 0̶9
    -  8 7 65
       3 7 44
   ```

10.
    ```
              14
         5 4̶9 12
      3,89̶6̶,5̶0̶ 2̶
    - 1,094,80 7
      2,801,69 5
    ```

11. Round 2847 to the nearest

    Ten: 2850 (284<u>7</u>; 7 is 5 or more)

    Hundred: 2800 (28<u>4</u>7; 4 is 4 or less)

    Thousand: 3000 (2<u>8</u>47; 8 is 5 or more)

12. Round 59,803 to the nearest

    Ten: 59,800 (59,80<u>3</u>; 0 is 4 or less)

    Hundred: 59,800 (59,8<u>0</u>3; 0 is 4 or less)

    Thousand: 60,000 (59,<u>8</u>03; 8 is 5 or more so 59 becomes 60

13. 3 × 9 × 7

    27 × 7   (3 × 9 = 27)

    189

14. $2 \times 8 \times 5$
    16×5 $(2 \times 8 = 16)$
        80

15. $9 \times 4 \times 6$
    36×6 $(9 \times 4 = 36)$
       216

16.   79
    $\times$ 8
    ─────
    632
    └─ Write 2, carry 7
    └─ Write 63

17.   58
    $\times$ 37
    ─────
     406
    174
    ─────
    2146

18.   845
    $\times$ 325
    ───────
      4 225
     16 90
    253 5
    ───────
    274,625

19.   1258
    $\times$ 420
    ───────
    25,160
    503 2
    ───────
    528,360

20. 530   53
    $\times$ 8 $\times$ 8
    ──────   4240   Attach 1 zero

21. 290   29
    $\times$ 50 $\times$ 5
    ──────   14,500   Attach 2 zeros

22.   389   389
    $\times$ 600 $\times$ 6
    ───────   233,400   Attach 2 zeros

23. We know how many boxes each lawn needs and the number of lawns. $80 \cdot 10 = 800$ boxes are required to do 80 lawns.

24. We know how many revolutions in one minute and the number of minutes. $50 \cdot 1800 = 90{,}000$ revolutions will be made in 50 minutes.

25. $\dfrac{56}{7} = \dfrac{7 \cdot 8}{7} = 8$

26.
$$\begin{array}{r} 158 \\ 9\overline{)1422} \\ \underline{9} \\ 52 \\ \underline{45} \\ 72 \\ \underline{72} \\ 0 \end{array}$$

27.
$$\begin{array}{r} 8\,975\ R2 \\ 3\overline{)26{,}927} \\ \underline{24} \\ 2\,9 \\ \underline{2\,7} \\ 22 \\ \underline{21} \\ 17 \\ \underline{15} \\ 2 \end{array}$$

28.
$$\begin{array}{r} 582 \\ 17\overline{)9894} \\ \underline{85} \\ 139 \\ \underline{136} \\ 34 \\ \underline{34} \\ 0 \end{array}$$

62 Chapter 3 Adding and Subtracting Fractions

29.
```
        4 710
   25)117,750
       100
        17 7
        17 5
           25
           25
           00
           00
            0
```

30.
```
          56 R 42
   286)16058
       1430
        1758
        1716
          42
```

31.
```
           32 R 166
   506)16,358
       15 18
        1 178
        1 012
          166
```

32. We need to divide the value of the coins into 5 equal parts.
```
        2 345
    5)11,725
      10
       1 7
       1 5
         22
         20
         25
         25
          0
```
Each child will receive $2345.

33. We need to find out how many times 16 will go into 9280.
```
        580
   16)9280
      80
       128
       128
        00
        00
         0
```
The vat can fill 580 cans.

34.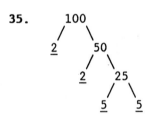

$30 = 2 \cdot 3 \cdot 5$

35.

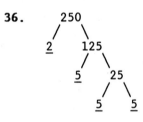

$250 = 2^2 \cdot 5^2$

36.

$250 = 2 \cdot 5^3$

37. $3^2 \cdot 2^4 = 3 \cdot 3 \cdot 2 \cdot 2 \cdot 2 \cdot 2 = 144$

38. $4^2 \cdot 5^2 = 4 \cdot 4 \cdot 5 \cdot 5 = 400$

39. $3^3 \cdot 5^2 = 3 \cdot 3 \cdot 3 \cdot 5 \cdot 5 = 675$

40. $\sqrt{9} = \sqrt{3 \cdot 3} = 3$

41. $\sqrt{64} = \sqrt{8 \cdot 8} = 8$

42. $\sqrt{225} = \sqrt{15 \cdot 15} = 15$

43. $8^2 - 8 \cdot 2 = 64 - 16$
    $= 48$

44. $\sqrt{25} + 5 \cdot 9 - 6 = 5 + 45 - 6$
    $= 50 - 6$
    $= 44$

45. $\frac{5}{6}$ proper
    Numerator is smaller than denominator

46. $\frac{3}{8}$ proper
    Numerator is smaller than denominator

47. $\frac{7}{4}$ improper
    Numerator is larger than denominator

48. $\frac{25}{40} = \frac{25 \div 5}{40 \div 5} = \frac{5}{8}$

49. $\frac{38}{50} = \frac{38 \div 2}{50 \div 2} = \frac{19}{25}$

50. $\frac{105}{300} = \frac{105 \div 15}{300 \div 15} = \frac{7}{20}$

51. $\frac{5}{\cancel{8}_2} \cdot \frac{\cancel{4}^1}{3} = \frac{5 \cdot 1}{2 \cdot 3} = \frac{5}{6}$

52. $\frac{9}{11} \cdot \frac{5}{\cancel{18}_2}^1 = \frac{1 \cdot 5}{11 \cdot 2} = \frac{5}{22}$

53. $25 \cdot \frac{3}{5} = \frac{\cancel{25}^5}{1} \cdot \frac{3}{\cancel{5}_1} = \frac{5 \cdot 3}{1 \cdot 1} = \frac{15}{1} = 15$

54. Area = length × width
    $= \frac{\cancel{4}^1}{5} \cdot \frac{3}{\cancel{8}_2} = \frac{3}{10}$

    Area = $\frac{3}{10}$ square inch.

55. $\frac{2}{5}$ of 10,000 goes to charity.

    $\frac{2}{\cancel{5}_1} \cdot \frac{\cancel{10000}^{2000}}{1} = \$4000$ to charity

    $\$10,000 - \$4000 = \$6000$ amount she has left.

    $\frac{2}{3}$ of 6000 goes to her son.

    $\frac{2}{\cancel{3}_1} \cdot \frac{\cancel{6000}^{2000}}{1} = \$4000$ goes to her son.

56. $\frac{5}{8} \div \frac{1}{4} = \frac{5}{\cancel{8}_2} \cdot \frac{\cancel{4}^1}{1} = \frac{5}{2} = 2\frac{1}{2}$

57. $\frac{25}{40} \div \frac{10}{35} = \frac{\cancel{25}^5}{\cancel{40}_8} \cdot \frac{35}{\cancel{10}_2} = \frac{35}{16} = 2\frac{3}{16}$

58. $6 \div \frac{3}{4} = \frac{\cancel{6}^2}{1} \cdot \frac{4}{\cancel{3}_1} = \frac{8}{1} = 8$

59. $\frac{1}{4} + \frac{1}{4} = \frac{1+1}{4} = \frac{2}{4} = \frac{1}{2}$

60. $\frac{15}{75} - \frac{10}{75} = \frac{15-10}{75} = \frac{5}{75} = \frac{1}{15}$

61. $\frac{1}{10} + \frac{2}{10} + \frac{3}{10} = \frac{1+2+3}{10} = \frac{6}{10} = \frac{3}{5}$

62.
```
2 | 25   30
3 | 25   15
5 | 25    5
5 |  5    1
     1    1
```
$150 = 2 \cdot 3 \cdot 5 \cdot 5$

63.
```
2 | 15   20   50
2 | 15   10   25
3 | 15    5   25
5 |  5    5   25
5 |  1    1    5
     1    1    1
```
$300 = 2 \cdot 2 \cdot 3 \cdot 5 \cdot 5$

64.
```
2 | 12   16   18
2 |  6    8    9
2 |  3    4    9
2 |  3    2    9
3 |  3    1    9
3 |  1    1    3
     1    1    1
```
$144 = 2 \cdot 2 \cdot 2 \cdot 2 \cdot 3 \cdot 3$

65. $\frac{5}{9} = \frac{}{72}$

$\frac{5 \cdot 8}{9 \cdot 8} = \frac{40}{72}$

66. $\frac{7}{12} = \frac{}{132}$

$\frac{7 \cdot 11}{12 \cdot 11} = \frac{77}{132}$

67. $\frac{9}{56} = \frac{}{168}$

$\frac{9 \cdot 3}{56 \cdot 3} = \frac{27}{168}$

68. $\frac{5}{7} = \frac{}{84}$

$\frac{5 \cdot 12}{7 \cdot 12} = \frac{60}{84}$

69. $\frac{2}{3} + \frac{1}{9} = \frac{6}{9} + \frac{1}{9} = \frac{6+1}{9} = \frac{7}{9}$

70. $\frac{5}{16} + \frac{1}{4} + \frac{3}{8} = \frac{5}{16} + \frac{4}{16} + \frac{6}{16}$

$= \frac{5+4+6}{16} = \frac{15}{16}$

71. $\frac{11}{15} - \frac{4}{25} = \frac{55}{75} - \frac{12}{75} = \frac{43}{75}$

72.
$2\frac{1}{4} = 2\frac{2}{8}$
$+ 3\frac{5}{8} = 3\frac{5}{8}$
$\phantom{+ 3\frac{5}{8} =\ } 5\frac{7}{8}$

73. $21\frac{7}{8} = 21\frac{21}{24}$
    $+ \ 4\frac{5}{12} = 4\frac{10}{24}$
    $\overline{\qquad\qquad\quad 25\frac{31}{24}} = 25 + 1\frac{7}{24} = 26\frac{7}{24}$

74. $5\frac{1}{8} = 5\frac{1}{8} = 4\frac{9}{8}$
    $- \ 2\frac{3}{4} = 2\frac{6}{8} = 2\frac{6}{8}$
    $\overline{\qquad\qquad\qquad\qquad\quad 2\frac{3}{8}}$

75–78. For Exercises 75 through 78, see the number line graphs in the back of your textbook.

NOTE: $\frac{5}{3} = 1\frac{2}{3}$; $\frac{10}{3} = 3\frac{1}{3}$

79. $\frac{7}{10} - \frac{37}{50}$
    $\frac{35}{50} < \frac{37}{50}$ Find common denominator

80. $\frac{19}{25} - \frac{23}{30}$
    $\frac{114}{150} < \frac{115}{150}$ Find common denominator

81. $\frac{7}{12} - \frac{11}{18}$
    $\frac{21}{36} < \frac{22}{36}$ Find common denominator

82. $\left(\frac{3}{8} - \frac{1}{3}\right) \cdot \frac{1}{2}$
    $= \left(\frac{9}{24} - \frac{8}{24}\right) \cdot \frac{1}{2}$
    $= \frac{1}{24} \cdot \frac{1}{2}$
    $= \frac{1}{48}$

83. $\frac{3}{4} \div \left(\frac{1}{3} + \frac{1}{2}\right)$
    $= \frac{3}{4} \div \left(\frac{2}{6} + \frac{3}{6}\right)$
    $= \frac{3}{4} \div \frac{5}{6}$
    $= \frac{3}{4} \cdot \frac{\overset{3}{6}}{\underset{2}{5}}$
    $= \frac{9}{10}$

84. $\frac{2}{3} + \left(\frac{7}{8}\right)^2 - \frac{1}{4}$
    $= \frac{2}{3} + \frac{49}{64} - \frac{1}{4}$
    $= \frac{128}{192} + \frac{147}{192} - \frac{1}{4}$
    $= \frac{275}{192} - \frac{1}{4}$
    $= \frac{275}{192} - \frac{48}{192}$
    $= \frac{227}{192}$
    $= 1\frac{35}{192}$

CHAPTER 4 DECIMALS

Section 4.1

3. .591

   hundredths: 9 (second digit to right of decimal point)

   thousandths: 1 (third digit to right of decimal point)

7. 27.658

   tens: 2 (second digit to left of decimal point)

   tenths: 6 (first digit to right of decimal point)

11. 6285.712

    thousands: 6 (fourth digit to left of decimal point)

    thousandths: 2 (third digit to right of decimal point)

15. .965

    9: tenths (first digit to right of decimal point)

    6: hundredths (second digit to right of decimal point)

    5: thousandths (third digit to right of decimal point)

19. $.7 = \frac{7}{10}$

23. $.45 = \frac{45}{100} = \frac{9}{20}$

27. $.41 = \frac{41}{100}$

31. $.405 = \frac{405}{1000} = \frac{81}{200}$

35. $.686 = \frac{686}{1000} = \frac{343}{500}$

39. $.64 = \frac{64}{100}$ = sixty-four hundredths

43. $12.4 = 12\frac{4}{10}$ = twelve and four tenths

47. Three and seven tenths = $3\frac{7}{10} = 3.7$

51. Four hundred twenty and three hundred eight thousandths

    $= 420\frac{308}{1000} = 420.308$

55. $4322.0531 = 4322\frac{531}{10,000}$

    = four thousand three hundred twenty-two and five hundred thirty-one ten-thousandths

59. 8235 = 8240 to the nearest ten

    8235 = 8200 to the nearest hundred

    8235 = 8000 to the nearest thousand

Section 4.2

3. Round to the nearest thousandth.

   965.498<u>3</u> 4 or less, 8 does not
   ↑   change. Drop digit to
              right of arrow.

   Answer: 965.498

7. Round to the nearest thousandth.

   27.905<u>6</u>1 5 or greater, 5 becomes
   ↑   6. Drop digits to right
               of arrow.

   Answer: 27.906

4.3 Addition of Decimals

11. Round to the nearest ten-thousandth.

    .0986<u>4</u>   4 or less, 6 does not change.
    ↑   Drop digit to right of arrow.
    Answer: .0986

15. Round to the nearest dollar.

    $139.<u>8</u>6   5 or greater, 9 becomes 10,
    ↑   which means 39 becomes 40.
    Answer: $140

19. 78.414 rounded to the nearest

    thousandths: 78.414 *Already written in thouandths*

    hundredths: 78.41 (4 is 4 or less, so 1 stays 1.)

    tenths: 78.4 (1 is 4 or less, so 4 stays 4.)

23. .0875 rounded to the nearest

    thousandths: .088 (5 is 5 or greater, so 7 becomes 8.)

    hundredths: .09 (7 is 5 or greater, so 8 becomes 9.)

    tenths: .1 (8 is 5 or greater, so 0 becomes 1.)

27. $10<u>9</u>.08 rounded to the nearest dollar is $109. (0 is 4 or less, 9 does not change.)

31. $70<u>5</u>.48 rounded to the nearest dollar is $705. (4 is 4 or less, 5 does not change.)

35. 614.7899153 rounded to the nearest

    millionth: 614.789915 (3 is 4 or less, so 5 stays 5.)

    hundred-thousandth:
        614.78992
        (5 is 5 or greater so 1 becomes 2.)

    ten-thousandth: 614.7899
        (1 is 4 or less, so 9 stays 9.)

39.   7929
       6076
     + 8218
     22,223

43.   81,976
           8
         785
      20,076
     + 7 208
     110,053

Section 4.3

3.   769.08
     406.15
    + 83.91
    1259.14

7. Line up decimal points.
    Attach 0's as needed.

     34.720
     19.812
    + 4.600
     59.132

11. Line up decimal points.
    Attach 0's as needed.

     39.765
    182.000
      4.719
      8.310
    + 5.900
    240.694

68   Chapter 4   Decimals

15. Write numbers in a column with decimal points lined up. Attach 0's as needed.

```
    27.650
    18.714
     9.749
  + 3.210
    59.323
```

19.
Problem	Estimate	Answer
482.70	483	482.70
16.92	17	16.92
+ 43.87	44	43.87
	544	543.49

The estimate and the answer are close.

23.
Problem	Estimate	Answer
62.81	63	62.810
539.9	540	539.900
5.629	6	5.629
	609	608.339

The estimate and the answer are close.

27.
Estimate	Answer	
5	4.5	days first week
6	6.25	days second week
+ 4	+ 3.74	days third week
15	14.49	days altogether

The estimate and the answer are close.

31. Estimate
```
    7542
     186
  + 186
    7914
```

Answer
```
  7542.1  miles on odometer at beginning
   186.4  miles to Visalia
+  186.4  miles back from Visalia
  7914.9  miles on odometer at end of
          trip
```

35. Add the length of each side.
```
     9.7100  yards
    14.0000  yards
  + 16.8044  yards
    40.5144  yards
```

39. Add the length of each side.
```
    32.0000  feet
    39.7100  feet
    46.9822  feet
    32.0000  feet
    25.8410  feet
    25.8410  feet
  + 25.8410  feet
   228.2152  feet
```

43.
```
          9
       6 10 18
       6̸ 7̸ 0̸ 8̸
     -   1  3  9
       6 5  6  9
```

Section 4.4

3. Line up the decimal points and subtract.
```
     10 11 12
     2̸ 1̸.2̸
   -   9  6.5
        1  5.7
```

7.
```
          13 10
       7  8̸  0̸ 13
      5 8.4̸  1̸  3̸
    - 2 5.8  4  7
      3 2.5  6  6
```

4.4 Subtraction of Decimals 69

11.
$$
\begin{array}{r}
7\ 15\ 6\ 12 \\
\cancel{8}\ \cancel{8}\ \cancel{7}\ \cancel{2}1 \\
-\ 3\ 7.6\ 80 \\
\hline
4\ 8.0\ 41
\end{array}
$$
(attach 0)

15. Line up decimal points. Attach 0's as necessary.
$$
\begin{array}{r}
6\ 12\ \ 8\ 10 \\
\cancel{7}\ \cancel{2}.\cancel{8}\cancel{9}\ \cancel{0} \\
-\ 2\ 7.65\ 4 \\
\hline
4\ 5.23\ 6
\end{array}
$$

19. Line up decimal points. Attach 0's as necessary.
$$
\begin{array}{r}
9\ 9 \\
7\ \cancel{10}\cancel{10}10 \\
1\cancel{8}.\cancel{0}\cancel{0}\cancel{0} \\
-\ \ \ .8\ 9\ 6 \\
\hline
17.\ 1\ 0\ 4
\end{array}
$$

23. Write numbers in a column with decimal points lined up. Attach 0's as necessary.
$$
\begin{array}{r}
17 \\
6\cancel{7}\ 14 \\
57\cancel{8}.\cancel{4}9 \\
-\ \ \ 69.80 \\
\hline
508.69
\end{array}
$$

27.

Problem	Estimate	Answer
		15 9
		7 $\cancel{5}$ $\cancel{10}$10
8.6	9	$\cancel{8}$.$\cancel{6}$ $\cancel{0}$ $\cancel{0}$
− 3.751	− 4	− 3.7 5 1
	5	4.8 4 9

The estimate and answer are close.

31.

Problem	Estimate	Answer
	16	16
	3$\cancel{8}$ 14	3$\cancel{8}$ 13
473.675	$\cancel{4}\cancel{7}$ $\cancel{4}$	$\cancel{4}\cancel{7}$ $\cancel{3}$.675
− 89.060	− 8 9	− 8 9.060
	38 5	38 4.615

The estimate and answer are close.

35. Place numbers in a column with decimal points lined up. Attach 0's where necessary.
$$
\begin{array}{r}
7\ 10\ 8\ 10 \\
114.\cancel{8}\ \cancel{9}\ \cancel{9}\ \cancel{9} \\
-\ 52.7\ 1\ 7\ 2 \\
\hline
62.0\ 9\ 1\ 8
\end{array}
$$

39. Estimate
$$
\begin{array}{r}
20 \\
-\ 9 \\
\hline
11
\end{array}
$$

$$
\begin{array}{rl}
9\ \ 9 & \\
1\ \cancel{10}\ \cancel{10}10 & \\
\$\cancel{2}\cancel{0}.\cancel{0}\cancel{0} & \text{given to clerk} \\
-\ \ \ \ 9.1\ 2 & \text{purchases} \\
\hline
\$1\ 0.8\ 8 & \text{change}
\end{array}
$$

43. Estimate
$$
\begin{array}{r}
31,562 \\
-\ 29,086 \\
\hline
2\ 476
\end{array}
$$

$$
\begin{array}{rl}
15 & \\
2\ 11\ 4\cancel{5}\ 11 & \\
\cancel{3}\cancel{1},\cancel{5}\cancel{6}\ \cancel{1}.9 & \text{odometer end of March} \\
-\ 2\ 9,08\ 6.1 & \text{odometer beginning of March} \\
\hline
2\ 47\ 5.8 & \text{miles driven}
\end{array}
$$

(The larger number must be on top to subtract).

47. Place numbers in a column with decimal points lined up. Attach 0's where necessary.
$$
\begin{array}{r}
10\ \ \ \ \ \ 9\ 9\ 9 \\
1\cancel{0}\ \ 10\ 3\ \cancel{10}\cancel{10}\cancel{10}10 \\
2\cancel{2}\cancel{1}.\cancel{0}\cancel{4}\cancel{0}\ \cancel{0}\ \cancel{0}\ \cancel{0} \\
-\ 218.\ 5\ 2\ 8\ 6\ 8\ 3 \\
\hline
2.\ 5\ 1\ 1\ 3\ 1\ 7
\end{array}
$$

Chapter 4 Decimals

51. Add the known sides and subtract the answer from the distance around.

```
           15 18 10
           4⁵ ⁸ ⁰ 11
  17.980   5⁶.⁹ 1 1
+ 19.006  - 36.9 8 6
  36.986   19.9 2 5 feet = length
                      of missing side.
```

55. Add the known lengths and subtract the answer from the total length.

```
                4 10
   3.569      9.9⁵ ⁰
 + 3.569    - 7.13 8
   7.138      2.81 2 inches = missing
                             length
```

59.
```
         3 789
   ×     2 205
        18 945
        00 00
       757 8
     7 578
   8,354,745
```

Section 4.5

3.
```
     47.62     2 decimal places
  ×   2.61   + 2 decimal places
     47 62     4 total decimal places
   2857 2
   9524      Estimate:    48
  124.2882                × 3
                         144
```

(When multiplying, the estimates will not be as accurate as the estimates when adding or subtracting)

7.
```
     280.9      1 decimal place
  ×   6.85    + 2 decimal places
    14045       3 total decimal places
    22472
   16854      Estimate:    281
  1924.165                 × 7
                          1967
```

11.
```
     21.7       1 decimal place
  ×   6.1     + 1 decimal place
     217        2 in product
    1302
   132.37
```

15.
```
     51.81      2 decimal places
  ×   .021    + 3 decimal places
     5181       5 in product
    10362
   1.08801
```

19.
```
     .0892      4 decimal places
  ×   .036    + 3 decimal places
     5352       7 in product
    2676
   .0032112   Two 0's must be attached
              to get 7 decimal places.
```

23.
```
     .0398      4 decimal places
  ×   .056    + 3 decimal places
     2388       7 in product
    1990
   .0022288   Two 0's must be attached
              to get 7 decimal places.
```

27.
```
     60.98      2 decimal places
  ×  .0018    + 4 decimal places
    48784       6 in product
    6098
   .109764
```

31.
```
     8.49       2 decimal places
  ×   .6      + 1 decimal place
     5.094      3 in product
```

35.
```
     .003       3 decimal places
  ×  .002     + 3 decimal places
    .000006     6 in product

              Zeros must be attached
              to get 6 decimal places.
```

4.6 Division of Decimals   71

39. $4.51 per hour   2 decimal places
     × 41.5 hours   + 1 decimal place
       2255   3 in product
       451
      1804
     187.165 — $187.17 (rounded)

43. $7.35 per hour   (3 decimal places)
     × 60.5 hours
       3675
      44100
     444.675 — $444.68 (rounded)

47. $1.72 per roll   (2 decimal places)
     × 15 rolls
       860
       172
      25.80 — $25.80

51.   6400 ball point pens
     × .022 each   (3 decimal places)
       12800
       12800
      140.800 — $140.80

55. $28.96 per month
     ×   24 months
       11584
        5792
       695.04 — $695.04

            Estimate: 29
                        × 24
                         116
                          58
                         696

59. Multiply the amount of fertilizer
     needed per acre by the number of
     acres to find the total number of
     gallons of fertilizer needed:

        158.25 acres
      ×   3.52 gallons per acre
         31650
         79125
         47475
        557.0400 — 557.04 gallons needed

Subtract the number of gallons needed
from the number of gallons in the
tank:

            9 9 9
         5 10̸10̸ 10̸10̸
         6̸0̸0̸.0̸0̸
        -5 5 7. 0 4
         4 2. 9 6 gallons remaining

              21 R33
63. 77)1650
         154
         110
          77
          33

              640 R8
67. 38)24,328
        22 8
        15 2
        15 2
           08
            0
            8

Section 4.6

            4.002
3. 5)20.010   Align decimal points
      20
       0 0   Attach zeros as needed
         0
         01
          0
          10
          10
           0

            .002
7. 4).008   Align decimal points
      0
      00
       0
       08
        8
        0

Chapter 4   Decimals

```
         11.69
11.  8)93.52      Align decimal points
       8
       13
        8
        5 5
        4 8
          72
          72
           0
```

```
           .004
15.  16).064
         0
         06
          0
          64
          64
           0
```

```
         .8469
19.  78)66.0590
       62 4
        3 65
        3 12
          539
          468
          710    Zero added to dividend and
          702    brought down
            8    Stop here   There are
                 enough digits to round
```

.8469   5 or greater.   6 becomes 7.
   ⁻   Drop the 9.

Answer: .847 Rounded to the nearest
              thousandths

23. .009)27̄

Move decimal point 3 places to the
right in both divisor and dividend.

```
            3 000
       9ᴧ)27,000ᴧ
          27
           0 0
             0
             00
              0
              00
               0
               0
```

Answer: 3000

```
            1.9180
27.  2.6ᴧ√4.9ᴧ8700    Move decimal point 1
       2 6            place to the right in
       2 3 8          both divisor and
       2 3 4          dividend
           47         Align decimal points
           26
           210        Zero added to dividend
           208
            20        Zero added to dividend
             0
            20        Stop here
```

Answer: 1.918

```
            .004
31.  .25ᴧ).00ᴧ100     Move decimal point 2
            0         places to the right
            10        in both divisor and
             0        dividend
            100       Align decimal points
            100
              0       Add zeros as needed
```

```
             3.036
35.  .025ᴧ).075ᴧ900   Move decimal point 3
            75        places to the right
             0 9      in both divisor and
             0 0      dividend
               90     Zero added to
               75     dividend
              150     Zero added to
              150     dividend
                0
```

Answer: 3.036

```
             56.
39.  .001ᴧ).056ᴧ      Move decimal point 3
                      places to the right
```

Answer: 56

4.6 Division of Decimals

43. 375.429 ÷ 12   Write as:

$$
\begin{array}{r}
31.2857 \\
12\overline{)375.4290} \\
\underline{36} \\
15 \\
\underline{12} \\
34 \\
\underline{24} \\
102 \\
\underline{96} \\
69 \\
\underline{60} \\
90 \quad \text{Zero added to dividend}\\
\underline{84} \\
6 \quad \text{Stop}
\end{array}
$$

Answer: 31.286   *Rounded to the nearest thousandth*

47.
$$
\begin{array}{r}
.03 \\
.034_\wedge\overline{).001_\wedge 02} \\
\underline{0} \\
1\ 02 \\
\underline{1\ 02} \\
0
\end{array}
$$

51. Divide. Total cost ÷ number of note pads = cost per pad.

$$
\begin{array}{r}
.422 \\
4\overline{)1.690} \\
\underline{1\ 6} \\
09 \\
\underline{8} \\
10
\end{array}
$$
Answer: $.42

55. Divide. Total cost ÷ number of pencils = cost per pencil.

$$
\begin{array}{r}
.051 \\
500\overline{)25.500} \\
\underline{25\ 00} \\
500 \\
\underline{500} \\
0
\end{array}
$$
Answer: $.05

59. Total cost ÷ number of bricks = cost per brick.

$$
\begin{array}{r}
.30 \\
619\overline{)185.70} \\
\underline{185\ 7} \\
00 \\
\underline{0} \\
0
\end{array}
$$
Answer: $.30

63.
$$
\begin{array}{r}
19.46 \quad \text{Amount paid each month}\\
\times\quad\ \ 21 \quad \text{Number of months}\\
\hline
19\ 46 \\
389\ 2 \\
\hline
\$408.66 \quad \text{Amount she owes}
\end{array}
$$

67. If 1 gross = 144 then 2 gross = 288. Total cost ÷ number of clips = cost per clip.

$$
\begin{array}{r}
.095 \\
288\overline{)27.360} \\
\underline{25\ 92} \\
1\ 440 \\
\underline{1\ 440} \\
0
\end{array}
$$
Answer: $.095 per paper clip

71. 9.152 ÷ 4.16 × 1.5
    = 2.2 × 1.5   *Divide first*
    = 3.3   *Multiply last*

75. 8.68 − 4.6 · 10.4 ÷ 6.4
    = 8.68 − 47.84 ÷ 6.4   *Multiply first*
    = 8.68 − 7.475   *Divide*
    = 1.205   *Subtract last*

Chapter 4 Decimals

79. Split the annual premium into four equal parts by dividing by 4.

$$\frac{234.5}{4 \overline{)938.0}} = \$234.50$$
$$\underline{8}$$
$$13$$
$$\underline{12}$$
$$18$$
$$\underline{16}$$
$$2\ 0 \quad \text{Add one zero to}$$
$$\underline{2\ 0} \quad \text{dividend and}$$
$$0 \quad \text{bring it down}$$

Add the additional $2.75 to find the amount of each payment.

$$\begin{aligned} 234.50 \\ +2.75 \\ \hline \$237.25 \end{aligned}$$

$237.25 is the amount of each quarterly payment.

83. $\frac{9}{5} \quad\underline{}\quad \frac{23}{15}$

15 is the least common denominator.

$$\frac{27}{15} > \frac{23}{15}$$

Section 4.7

3. $\frac{1}{3}$
$$\begin{array}{r} .3333 \\ 3\overline{)1.0000} \\ \underline{9} \\ 10 \\ \underline{9} \\ 10 \\ \underline{9} \\ 10 \\ \underline{9} \\ 1 \end{array}$$ Stop here   We have enough digits to round

Answer: .333 Rounded

7. $5\frac{1}{8} = \frac{41}{8}$

$$\begin{array}{r} 5.125 \\ 8\overline{)41.000} \\ \underline{40} \\ 1\ 0 \\ \underline{8} \\ 20 \\ \underline{16} \\ 40 \\ \underline{40} \\ 0 \end{array}$$ 3 zeros added to dividend

Answer: 5.125

11. $\frac{2}{3}$
$$\begin{array}{r} .6666 \\ 3\overline{)2.0000} \\ \underline{1\ 8} \\ 20 \\ \underline{18} \\ 20 \\ \underline{18} \\ 20 \\ \underline{18} \\ 2 \end{array}$$ 4 zeros added to dividend

Stop

Answer: .667 Rounded

15. $3\frac{22}{25} = \frac{97}{25}$

$$\begin{array}{r} 3.88 \\ 25\overline{)97.00} \\ \underline{75} \\ 22\ 0 \\ \underline{20\ 0} \\ 2\ 00 \\ \underline{2\ 00} \\ 0 \end{array}$$ 2 zeros added to dividend

Answer: 3.88

4.7 Writing Fractions as Decimals 75

19. $\dfrac{6}{7}$

$$\begin{array}{r} .8571 \\ 7\overline{)6.0000} \\ \underline{5\ 6} \\ 40 \\ \underline{35} \\ 50 \\ \underline{49} \\ 10 \\ \underline{7} \\ 3 \end{array}$$ 4 zeros added to dividend

Stop

Answer: .857   Rounded

23. $15\dfrac{17}{36} = \dfrac{557}{36}$

$$\begin{array}{r} 15.4722 \\ 36\overline{)557.0000} \\ \underline{36} \\ 197 \\ \underline{180} \\ 17\ 0 \\ \underline{14\ 4} \\ 2\ 60 \\ \underline{2\ 52} \\ 80 \\ \underline{72} \\ 80 \\ \underline{72} \\ 8 \end{array}$$ 4 zeros added to dividend

Stop

Answer: 15.472   Rounded

27. $.4 = \dfrac{4}{10} = \dfrac{2}{5}$

31. $.875 = \dfrac{875}{1000} = \dfrac{7}{8}$

35. $\dfrac{7}{20}$

$$\begin{array}{r} .35 \\ 20\overline{)7.00} \\ \underline{6\ 0} \\ 1\ 00 \\ \underline{1\ 00} \\ 0 \end{array}$$

Answer: .35

39. $\dfrac{5}{6}$

$$\begin{array}{r} .8333 \\ 6\overline{)5.0000} \\ \underline{4\ 8} \\ 20 \\ \underline{18} \\ 20 \\ \underline{18} \\ 20 \\ \underline{18} \\ 2 \end{array}$$ 4 zeros added

Stop

$\dfrac{5}{6} = .833$ Rounded

43. $.15 = \dfrac{15}{100} = \dfrac{3}{20}$

47. $\dfrac{3}{8}$ ___ .38 $\dfrac{3}{8} = .375$

Writing .38 as .380 and ignoring the decimal points we see that .375 is smaller than .380.

Answer: $\dfrac{3}{8} < .38$

51. $\dfrac{5}{8}$ ___ .60 $\dfrac{5}{8} = .625$

Writing .60 as .600 and ignoring the decimal points we see that .600 is smaller than .625.

Answer: $\dfrac{5}{8} > .60$

55. $\dfrac{2}{3}$ ___ .64 $\dfrac{2}{3} = .667$

Writing .64 as .640 we see that .640 is smaller than .667.

Answer: $\dfrac{2}{3} > .64$

Chapter 4 Decimals

59. $\frac{3}{5}$, .062, .55 $\frac{3}{5}$ = .6

Writing each decimal with 3 places
.600, .062, .550 we see that .062 is the smallest. Circle .062.

63. .0909, .091, $\frac{1}{11}$ $\frac{1}{11}$ = .090909

Write each decimal with 6 places
.090900, .091000, .090909
We see that .090900 is the smallest.
Circle .0909.

67. .17, $\frac{1}{6}$, $\frac{2}{13}$ $\frac{1}{6}$ = .167   (rounded)

$\frac{2}{13}$ = .154   (rounded)

Write each decimal with 3 places.
.170, .167, .154
We see that .154 is the smallest.
Circle $\frac{2}{13}$.

71. .54, .5455, .5399

Write each decimal with 4 places then arrange in order, smallest to greatest .5399, .54, .5455

75. .628,   .62812,   .609, $\frac{5}{8}$

.62800, .62812, .60900, .62500
Write each as a 5 place decimal and compare

Answer: .609, $\frac{5}{8}$, .628, .62812

79. .043, .051, .506, $\frac{1}{20}$

.043, .051, .506, .050
Write each as a 3 place decimal and compare

Answer: .043, $\frac{1}{20}$, .051, .506

83. $\frac{6}{11}$, $\frac{5}{9}$, $\frac{4}{7}$, .571

Write each as a decimal with 4 places. .5455, .5556, .5714, .5710
Arrange in order, smallest to greatest.
.5455, .5556, .5710, .5714

Answer: $\frac{6}{11}$, $\frac{5}{9}$, .571, $\frac{4}{7}$

87. $\frac{3}{16}$, $\frac{1}{4}$, $\frac{1}{5}$,   .188

.1875,   .2500,   .2000,   .1880
Write each as a 4 place decimal and compare

Answer: $\frac{3}{16}$, .188, $\frac{1}{5}$, $\frac{1}{4}$

91. $\frac{60}{80} = \frac{60 \div 20}{80 \div 20} = \frac{3}{4}$

Chapter 4 Review Exercises

1. 6.58
   └─ 8 is in the hundredths place.
   └─ 5 is in the tenths place.

2. .7853
   └─ 3 is in the ten-thousandths place.
   └─ 7 is in the tenths place.

3. 435.621
   — 1 is in the thousandths place.
   — 6 is in the tenths place.

4. 896.503
   — 0 is in the hundredths place.
   — 5 is in the tenths place.

5. 620.738
   — 8 is in the thousandths place.
   — 7 is in the tenths place.

6. $.9 = \frac{9}{10}$

7. $.75 = \frac{75}{100} = \frac{3}{4}$

8. $.03 = \frac{3}{100}$

9. $.875 = \frac{875}{1000} = \frac{7}{8}$

10. $.6158 = \frac{6158}{10,000} = \frac{3079}{5000}$

11. $.8895 = \frac{8895}{10,000} = \frac{1779}{2000}$

12. .8 = eight tenths

13. .45 = forty-five hundredths

14. 12.87 = twelve and eighty-seven hundredths

15. 335.708 = three hundred thirty-five and seven hundred eight thousandths

16. 42.105 = forty-two and one hundred five thousandths

17. 275.6<u>3</u>5   4 or less, 6 does not change.
    ↑   Drop digits to right of arrow.

    Answer: 275.6   *Rounded to the nearest tenth*

18. 72.78<u>9</u>   5 or greater, 8 becomes 9.
    ↑   Drop digit to right of arrow.

    Answer: 72.79   *Rounded to the nearest hundredth*

19. 896.2<u>5</u>3   4 or less, 5 does not change.
    ↑   Drop digit to right of arrow.

    Answer: 896.25   *Rounded to the nearest hundredth*

20. .023<u>5</u>   5 or greater, 3 becomes 4.
    ↑   Drop digit to right of arrow.

    Answer: .024   *Rounded to the nearest thousandth*

21. 87.<u>4</u>798   4 or less, 7 does not change.
    ↑   Drop digits to the right of arrow.

    Answer: 87   *Rounded to the nearest one*

22. $15.<u>8</u>3   5 or greater, 5 becomes 6.
    Answer: $16

Chapter 4 Decimals

23. $81.<u>5</u>1   5 or greater, 1 becomes 2.
    Answer: $82

24. $17,625.<u>7</u>9   5 or greater, 5 becomes 6.
    Answer: $17,626

25. $78.<u>5</u>8   5 or greater
    Answer: $79

26. $37.<u>2</u>8   4 or less
    Answer: $37

27. $19.<u>2</u>0   4 or less
    Answer: $19

28.
Problem	Estimate	Answer
78.56	79	78.56
22.15	22	22.15
+ 39.68	+ 40	+ 39.68
	141	140.39

The answer and the estimate are close.

29.
Problem	Estimate	Answer
5.8	6	5.80
23.96	24	23.96
+ 15.09	+ 15	+ 15.09
	45	44.85

The answer and the estimate are close.

30.
Problem	Estimate	Answer
3.58	4	3.58
7.9	8	7.90
5.7	6	5.70
8.65	9	8.65
+ 20.4	+ 20	+ 20.40
	47	46.23

The answer and the estimate are close.

31.
Problem	Estimate	Answer
75.6	76	75.600
1.63	2	1.630
22.045	22	22.045
1.88	2	1.880
+ 33.7	+ 34	+ 33.700
	136	134.855

The answer and the estimate are close.

32.
Problem	Estimate	Answer
45.6	46	45.60
7.09	7	7.09
5.63	6	5.63
+ 78.09	+ 78	+ 78.09
	137	136.41

The answer and the estimate are close.

33. | Problem | Estimate | Answer |
|---|---|---|
| 11.206 | 11 | 11.206 |
| 50.3 | 50 | 50.300 |
| 77.8 | 78 | 77.800 |
| 9.05 | 9 | 9.050 |
| + 11.7 | + 12 | + 11.700 |
| | 160 | 160.056 |

The answer and the estimate are close.

34. | Problem | Estimate | Answer |
|---|---|---|
| 28.6 | 29 | 28.6 |
| − 17.4 | − 17 | − 17.4 |
| | 12 | 11.2 |

The answer and the estimate are close.

35. | Problem | Estimate | Answer |
|---|---|---|
| | | 7 15 |
| 28.5 | 29 | 28.$\cancel{5}$ |
| − 17.8 | − 18 | − 17.8 |
| | 11 | 10.7 |

The answer and the estimate are close.

36. | Problem | Estimate | Answer |
|---|---|---|
| | | 4 1 6 |
| 36.356 | 36 | 36.3$\cancel{5}\cancel{6}$ |
| − 25.348 | − 25 | − 25.348 |
| | 11 | 11.008 |

The answer and the estimate are close.

37. | Problem | Estimate | Answer |
|---|---|---|
| | | 7 17 |
| 8.731 | 9 | 8.$\cancel{7}$31 |
| − 7.8 | − 8 | − 7.800 |
| | 1 | .931 |

The answer and the estimate are close.

38.
```
      9
   2 10 10
   3 0. 0    hours to be donated
 − 1 9. 6    hours already worked
   1 0. 4    hours to work
```

39.
```
           14
      1 11 4 4 13
   $ 2 1 5. 5 3    cash for check
   −     4 3. 8 9  amount for groceries
   $ 1 7 1. 6 4    change
```

40.
```
   $ 1.18    toothpaste
     5.83    gift
   + 15.94   humidifier
   $22.95    total spent
```

41.
```
     2.36   Monday
     3.58   Tuesday
     1.78   Wednesday
     3.90   Thursday
   + 5.35   Friday
    16.97   total miles jogged
```

42.
```
      .312    3 decimal places
   ×  5.6    +1 decimal place
     1872    4 decimal places in
     1560      product
    1.7472
```

43.
```
     6.138    3 decimal places
   × .037    +3 decimal places
    42966    6 decimal places in
    18414      product
    .227106
```

80   Chapter 4   Decimals

44.   42.09   2 decimal places
     × .023   +3 decimal places
     ──────
       12627   5 decimal places
        8418   in product
     ──────
     .96807

45.   5.6   1 decimal place
     × .02 +2 decimal places
     ─────
      .112   3 decimal places
              in product

46. $2.46 per pound   2 decimal places
     ×   41 pounds   in product
     ──────
        246
       984
     ──────
     $100.86 total cost

47. $.088 per pack 3 decimal places
     × 48   packs   in product
     ──────
       704
      352
     ──────
     4.224 = $4.22 total cost (rounded)

48. $3.57 wages per hour 2 decimal
     ×   36 hours   places in
     ──────   product
      2142
     1071
     ──────
     $128.52 total earnings

49. $26.48 payment per month 2 decimal
     ×   18 months   places in
     ──────   product
     21184
      2648
     ──────
     $476.64 total of payments

50.   14.5
     3)43.5   Align decimal points
       3
      ──
      13
      12
      ──
       15
       15
       ──
        0

     Answer: 14.5

51.   4.7153
     7.56ʌ)35.64ʌ8000 Move decimal point
          30 24   2 places to the
           5 40 8   right in both divi-
           5 29 2   sor and dividend
             11 60   Zero added to
              7 56   dividend
              4 040   Zero added to
              3 780   dividend
               2600   Zero added to
               2268   dividend
                332   Stop

     Answer: 4.715

52.   155 00.
     .05ʌ)775.00ʌ In order to move
         5   decimal point 2
        ──   places to the right,
        27   two zeros must be
        25   added to dividend
        ──
         25
         25
         ──
          0

     Answer: 15,500

53.   .4
     .0012ʌ).0004ʌ8 Move decimal 4 places
            4 8   to the right in both
            ──   dividend and divisor
             0

     Answer: .4

54. Purchase price ÷ price per share
    = shares purchased.

    $$33.75\overline{\smash{)}2970.00}$$

    ```
              88
    33.75.)2970.00.
           2700 0
            270 00
            270 00
                 0
    ```

    Answer: 88 shares can be purchased.

55. Earnings ÷ hours worked = earnings per hour.

    ```
              6.4
    44.75.)286.40.0
           268 50
            17 90 0
            17 90 0
                  0
    ```

    Answer: Earnings per hour are $6.40.

56. $3.5^2 + 8.7 - 1.95$
    = 12.25 + 8.7 − 1.95   *Exponent first*
    = 20.95 − 1.95   *Add*
    = 19   *Subtract*

57. 3.16 ÷ 3.95 − .33
    = .8 − .33   *Divide first*
    = .47   *Subtract*

58. $\frac{4}{5}$
    ```
         .8
    5)4.0    1 zero added to dividend
    ```
    Answer: .8

59. $\frac{16}{25}$
    ```
         .64
    25)16.00    2 zeros added
       15 0     to dividend
        1 00
        1 00
           0
    ```
    Answer: .64

60. $\frac{14}{21}$
    ```
         .6666
    21)14.0000    4 zeros added
       12 6       to dividend
        1 40
        1 26
          140
          126
           140
           126
            14
    ```
    Answer: .667 (rounded)

61. $\frac{1}{9}$
    ```
         .1111
    9)1.0000    4 zeros added
      9         to dividend
      10
       9
       10
        9
        10
         9
         1
    ```
    Answer: .111 (rounded)

62. Write each decimal with 5 places.
    .68000, .68210, .67295
    Arrange in order, smallest to greatest.
    .67295, .68000, .68210
    Answer: .67295, .68, .6821

63. Write each decimal with 4 places.
    .2150, .2200, .2090, .2102
    Arrange in order, smallest to greatest.
    .2090, .2102, .2150, .2200
    Answer: .209, .2102, .215, .22

82   Chapter 4   Decimals

64. .17, $\frac{3}{20}$, $\frac{1}{6}$, .159

Write each as a decimal with 3 places.

.170, .150, .167, .159

Arrange in order, smallest to greatest.

.150, .159, .167, .170

Answer: $\frac{3}{20}$, .159, $\frac{1}{6}$, .17

65.
```
Problem      Estimate     Answer
 72.11         72          72.110
  5.06          5           5.060
 31.673        32          31.673
  4.08          4           4.080
+ 9.7         +10         + 9.700
                123         122.623
```

The estimate and the answer are close.

66.
```
  72.8      1 decimal place
×  3.5     +1 decimal place
  3540      2 decimal places
  2184         product
 254.80
```

67.
```
           35 83.2608
  .46ʌ)1648.30ʌ0000     Add 5 zeros to
       138              dividend and move
       268              decimal 2 places
       230              to the right in
       383              both divisor and
       368              dividend
       150
       138
       120
        92
        28 0
        27 6
           400
           368
            32   Stop
```

Answer: 3583.261   (rounded)

68.
```
Problem    Estimate     Answer
                         9 11 9
                        6 10 ⃫10̸10
 70.2         70        7̸0̸.2̸0̸0̸
-35.668      -36       -3 5. 6 6 8
              34        3 4. 5 3 2
```

The answer and the estimate are close.

69.
```
  34.28      2 decimal places
×   .08     +2 decimal places
  2.7424     4 decimal places
                in product
```

Answer: 2.742   (rounded)

70.
```
              .1214
   .387ʌ).047ʌ0000     Move decimal 3
         38 7          places to the right
          8 30         in both divisor and
          7 74         dividend and add 4
            560        zeros to dividend
            387
           1730
           1548
            182   Stop
```

Answer: .121   (rounded)

71.
```
Problem    Estimate     Answer
 72.105       72         72.105
  8.2          8          8.200
+95.37       +95        +95.370
             175         175.675
```

The answer and the estimate are close.

72.
```
         9.04
   9)81.36       Bring decimal point
      81         straight up
       0 3
       0
         36
         36
          0
```

Answer: 9.04

73.
```
    Problem      Estimate      Answer
                                  10
                                1 ⁹̷ 10
    21.059          21        2̷ 1̷ . ⁰̷59
   -  3.8         -  4        -  3. 800
                    17         1 7. 259
```

The answer and the estimate are close.

74.
```
      .38       2 decimal places
    × .07     + 2 decimal places
      266       4 decimal places
       00         in product
     .0266   One zero must be added
```

Answer: .027   (rounded)

75.
```
      1.60      2 decimal places
    × .508    + 3 decimal places
      1280      5 decimal places in
      8000        product
     .81280
```

Answer: .813   (rounded)

76.
```
                 80.8715
        .218ᴧ)17.630ᴧ0000    Add 5 zeros to
               17 44         dividend and move
                1900         decimal 3 places
                1744         to the right in
                 1560        both divisor and
                 1526        dividend
                   340
                   218
                  1220
                  1090
                   130   Stop
```

Answer: 80.872   (rounded)

77. $(5.6 - 1.22) \cdot 4.8 \div 3.15$

   $= 4.38 \cdot 4.8 \div 3.15$ Subtract in parentheses first

   $= 21.024 \div 3.15$ Multiply

   $= 6.674$ (rounded)

78. $2^3 + 1.3 - 5.6$

   $= 8 + 1.3 - 5.6$ Exponent first

   $= 9.3 - 5.6$ Add

   $= 3.7$ Subtract

79. $\frac{2}{5}$
```
           .4
        5)2.0   1 zero added
                   to dividend
```
Answer: .4

80. $\frac{11}{20}$
```
            .55
        20)11.00   2 zeros added
           10 0       to dividend
            1 00
            1 00
               0
```
Answer: .55

81. $\frac{3}{10}$
```
           .3
        10)3.0   1 zero added
                    to dividend
```
Answer: .3

82. $\frac{1}{7}$
```
           .1428
        7)1.0000   4 zeros added
          7           to dividend
          30
          28
           20
           14
            60
            56
             4
```
Answer: .143   (rounded)

83. $\frac{3}{8}$, .381, .3749, $\frac{2}{5}$

Write each as a decimal with 4 places.

   .3750, .3810, .3749, .4000

Arrange in order, smallest to greatest.

.3749, .3750, .3810, .4000

Answer: .3749, $\frac{3}{8}$, .381, $\frac{2}{5}$

84. .348, $\frac{7}{20}$, $\frac{8}{23}$, .375

Write each as a decimal with 4 places.

.3480, .3500, .3478, .3750

Arrange in order, smallest to greatest.

.3478, .3480, .3500, .3750

Answer: $\frac{8}{23}$, .348, $\frac{7}{20}$, .375

85. miles ÷ gallons = miles per gallon

```
        1 5.52
18.4₌)285.6₌00
      184
      101 6
       92 0
        9 60
        9 20
          400
          368
           32
```

Answer: 15.5 miles per gallon

86.   412.7 gallons
  × 1.09 dollars per gallon
  37 143
  00 00
  412 7
  449.843 = $449.84 total cost
               (rounded)

87. Amount of loan ÷ monthly payment = months to repay.

```
            18.
174.25₌)3136.50₌
        1742 5
        1394 00
        1394 00
              0
```

Answer: 18 months are needed to repay the loan.

88. $38.52 credit
  + 14.24 credit
  $52.76 total credits

  $ 9.36 charge
    17.29 charge
    85.82 charge
    23.59 charge
    16.73 charge
  + 46.71 charge
  $199.50 total charges

The balance now due is the amount of the total charges minus the amount of the total credits.

  $199.50
  − 52.76
  $146.74 is now due.

89. $ 89.14 spent
   14.37 spent
    .75 spent
   3.68 spent
 $107.94 total amount spent

Subtract the total amount spent from what she started with to find out how much is left.

  $125.50 amount started with
  − 107.94 amount spent
  $ 17.56 remains

90. $19.95   daily charge
     ×   4   days
     $79.80   total daily charge

     $ 7.80   daily insurance
     ×   4   days
     $31.20   total insurance charge

        652   miles
     × .15   dollars per mile
     $97.80   total milege charge

Now find the grand total.

     $79.80   daily charge
      31.20   insurance
     + 97.80 mileage charge
     $208.80 total cost of the rental

Chapter 4   Test

1. $.937 = \dfrac{937}{1000}$

2. $.3053 = \dfrac{3053}{10{,}000}$

3. $.875 = \dfrac{875}{1000} = \dfrac{7}{8}$

4. $.0075 = \dfrac{75}{10{,}000} = \dfrac{3}{400}$

5. 725.60895   5 or greater, 8 becomes 9.
         ↑   Drop digits to right
                of arrow.
    Answer: 725.609

6. .705149   4 or less, 1 does not
         ↑   change. Drop digits to
              right of arrow.
    Answer: .7051

7. $611.49   4 or less, 1 does not
        ↑   change.
    one dollar
    Answer: $611

8. $7859.51 5 or greater, 9 becomes 10
        ↑   which means 59 becomes 60.
    one dollar
    Answer: $7860

9. Write numbers in a column   7.6059
    with decimal points lined   82.0128
    up. Add 0's as necessary.   + 32.5900
                                  122.2087

10.   53.182
        4.631
      782.052
         .031
      839.896

11. Place largest number on
    top and line up decimal
    points. Add 0's as
    necessary.

    $10\ 9$
    $8\ \cancel{0}\ \cancel{10}10$
    $7\cancel{9}.\cancel{X}\ \cancel{8}\ \cancel{0}$
    $-\ 23.6\ 0\ 2$
    $\ \ \ 55.4\ 9\ 8$

12. 1710
     $6\ \cancel{7}\ \cancel{0}15$
     $66\cancel{7}.\cancel{8}\ \cancel{X}\ \cancel{8}$
     $-\ \ \ \ .9\ 9\ 6$
     $666.8\ 1\ 9$

13.   45.79   2 decimal places
     × .03   + 2 decimal places
     1.3737   4 total decimal places

86 Chapter 4 Decimals

14.
```
    .0069      4 decimal places
  × .007     + 3 decimal places
  .0000483    7 total decimal places
              4 0's must be added
```

15. $\dfrac{96}{.015} = 96 \div .015$

```
          64 00.
.015 ) 960 00.    Move decimal point 3
       90         places to the right
       ─          in both divisor and
        60        dividend
        60
        ──
         00
          0
         ──
          00
           0
          ──
           0
```

Answer: 6400

16.
```
            6 05.03
.15 ) 90.75.45    Move decimal point 2
     90           places to the right
     ──           in both divisor and
      0 7         dividend
        0
        ──
        75        Align decimal points
        75
        ──
         04
          0
         ──
          45
          45
          ──
           0
```

17. Write each decimal with 4 places.
   .5080, .5160, .5108
   Write smallest to greatest (ignoring decimal).
   .5080, .5108, .5160
   Answer: .508, .5108, .516

18. Write each as decimal with 4 places.
   .4400, .4510, .4500, .4606
   Write smallest to greatest.
   .4400, .4500, .4510, .4606
   Answer: .44, $\dfrac{9}{20}$, .451, .4606

19. Write each as decimal with 4 places.
   .6020, .6000, .5983, .6667
   Write smallest to greatest.
   .5983, .6000, .6020, .6667
   Answer: .5983, $\dfrac{3}{5}$, .602, $\dfrac{2}{3}$

20. $6.3^2 - 5.9 + 3.4 \cdot (.5)$
   $= 39.69 - 5.9 + 3.4 \cdot (.5)$ Exponent first
   $= 39.69 - 5.9 + 1.7$ Multiply
   $= 33.79 + 1.7$ Subtract left to right
   $= 35.49$ Add

21. Add all the checks together and subtract the answer from the balance in her account.

```
   $    5.73        $271.15
       17.92       - 206.39
       29.56       ───────
   +  153.18        $ 64.76
      ──────
      $206.39
```

Answer: After writing the checks, Jennifer had a balance of $64.76.

22.
```
      38.25   gallons in one drum
   ×  10.6   drums
   ──────
      22 950
      00 00
     382 5
   ───────
     405.450 = 405.45 gallons
```

23. Divide the total miles by the number of gallons to get miles per gallon.

```
          2 7.71
  12.7ˬ)352.0ˬ00    Attach 3 zeros and
        254         move decimal point
         980        one place to the
         889        right in both divisor
          910       and dividend
          889
           210
           127
            83      Stop
```

Answer: 27.7 miles per gallon (rounded)

24. Divide the cost by the number of pounds to find cost per pound.

```
           2.78
  12.5ˬ)34.7ˬ50     Attach 1 zero and move
        25 0        decimal point 1 place
         9 7 5      to right in both
         8 7 5      divisor and dividend
         1 0 00
         1 0 00
               0
```

Answer: $2.78 per pound

25. Multiply.

```
           32 days
         x 4.6 therms each day
          192
          128
          147.2
```

Answer: 147.2 therms were used.

CHAPTER 5 RATIO AND PROPORTION

Section 5.1

3. $\dfrac{75}{100} = \dfrac{3}{4}$

7. $\dfrac{80 \text{ miles}}{50 \text{ miles}} = \dfrac{8}{5}$

11. $\dfrac{6\frac{1}{2}}{4} = \dfrac{\frac{13}{2}}{\frac{4}{1}} = \dfrac{13}{1} \div \dfrac{4}{1} = \dfrac{13}{2} \cdot \dfrac{1}{4} = \dfrac{13}{8}$

15. $\dfrac{1\frac{1}{4}}{1\frac{1}{2}} = \dfrac{\frac{5}{4}}{\frac{3}{2}} = \dfrac{5}{4} \div \dfrac{3}{2} = \dfrac{5}{\cancel{4}} \cdot \dfrac{\cancel{2}}{3} = \dfrac{5}{6}$

19. $\dfrac{5 \text{ minutes}}{1 \text{ hour}} = \dfrac{5 \text{ minutes}}{1 \cdot 60 \text{ minutes}}$
    $= \dfrac{5}{60} = \dfrac{1}{12}$

23. $\dfrac{1\frac{1}{2} \text{ days}}{1 \text{ week}} = \dfrac{1\frac{1}{2} \text{ days}}{1 \cdot 7 \text{ days}} = \dfrac{1\frac{1}{2}}{7}$
    $= \dfrac{\frac{3}{2}}{\frac{7}{1}} = \dfrac{3}{2} \div \dfrac{7}{1}$
    $= \dfrac{3}{2} \cdot \dfrac{1}{7} = \dfrac{3}{14}$

27. $\dfrac{2\frac{2}{3} \text{ cubic feet}}{6 \text{ cubic feet}} = \dfrac{\frac{8}{3}}{\frac{6}{1}} = \dfrac{8}{3} \div \dfrac{6}{1}$
    $= \dfrac{\cancel{8}^{4}}{3} \cdot \dfrac{1}{\cancel{6}_{3}} = \dfrac{4}{9}$

31. $\dfrac{\text{longest}}{\text{shortest}} = \dfrac{7}{5}$

35. $\dfrac{\text{longest}}{\text{shortest}} = \dfrac{9\frac{1}{2}}{4\frac{1}{4}} = \dfrac{\frac{19}{2}}{\frac{17}{4}} = \dfrac{19}{2} \div \dfrac{17}{4}$
    $= \dfrac{19}{\cancel{2}_{1}} \cdot \dfrac{\cancel{4}^{2}}{17} = \dfrac{38}{17}$

39. $\begin{array}{r} \$9.90 \\ -\ 6.60 \\ \hline \$3.30 \end{array}$ new price
    original price
    increase in price

    $\dfrac{\text{increase}}{\text{original}} = \dfrac{\$3.30}{\$6.60}$
    $= \dfrac{3.30 \div 3.30}{6.60 \div 3.30}$
    $= \dfrac{1}{2}$

43. $\begin{array}{r} 32.81 \\ \times\ 16.83 \\ \hline 98\ 43 \\ 26\ 24\ 8 \\ 196\ 86 \\ 328\ 1 \\ \hline 552.19\ 23 \end{array}$ 2 decimal places
    + 2 decimal places
    4 decimal places
    in product

    Answer: 552.192 (rounded)

Section 5.2

3. $\dfrac{250 \text{ yards}}{50 \text{ seconds}} = \dfrac{5 \text{ yards}}{1 \text{ second}}$

7. $\dfrac{35 \text{ gallons}}{5 \text{ hours}} = \dfrac{7 \text{ gallons}}{1 \text{ hour}}$

11. $\dfrac{72 \text{ miles}}{4 \text{ gallons}} = \dfrac{18 \text{ miles}}{1 \text{ gallon}}$

15. $\dfrac{1260 \text{ dollars}}{14 \text{ days}} = \dfrac{90 \text{ dollars}}{1 \text{ day}}$
    $= \dfrac{90 \text{ dollars}}{\text{day}}$

19. $\dfrac{311.08 \text{ miles}}{15.4 \text{ gallons}} = \dfrac{20.2 \text{ miles}}{1 \text{ gallon}}$

$\phantom{\dfrac{311.08 \text{ miles}}{15.4 \text{ gallons}}} = \dfrac{20.2 \text{ miles}}{\text{gallon}}$

23. $\dfrac{413.20 \text{ dollars}}{4 \text{ days}} = \dfrac{103.3 \text{ dollars}}{1 \text{ day}}$

$\phantom{\dfrac{413.20 \text{ dollars}}{4 \text{ days}}} = \dfrac{103.3 \text{ dollars}}{\text{day}}$

27. Find the unit price for each and compare.

Size	Unit Cost	
10 oz.	$\dfrac{\$1.22}{10 \text{ ounces}}$	= $.122 per ounce
15 oz.	$\dfrac{\$1.66}{15 \text{ ounces}}$	= $.111 per ounce
20 oz.	$\dfrac{\$2.11}{20 \text{ ounces}}$	= $.106 per ounce

The best buy is the 20 ounce box.

31. 20 pounds serves 15 people.

$\dfrac{20 \text{ pounds}}{15 \text{ people}}$ *Divide 20 by 15*

$\dfrac{\overset{4}{\cancel{20}}}{\underset{3}{\cancel{15}}} = 1\dfrac{1}{3} \dfrac{\text{pounds}}{\text{person}}$

35. $\dfrac{1725 \text{ dollars}}{150 \text{ shares}} = \dfrac{11.5 \text{ dollars}}{\text{share}}$

or $11.50 per share

39. 4.6 yards cost $51.75. Find cost (dollars) per yard.

$\dfrac{51.75 \text{ dollars} \div 4.6}{4.6 \text{ yards} \div 4.6} = \dfrac{11.25 \text{ dollars}}{1 \text{ yard}}$

$\phantom{\dfrac{51.75 \text{ dollars} \div 4.6}{4.6 \text{ yards} \div 4.6}} = \dfrac{\$11.25}{\text{yard}}$

43. $18\dfrac{2}{5} \cdot 30$

$= \dfrac{92}{\cancel{5}} \cdot \dfrac{\overset{6}{\cancel{30}}}{1}$

$= \dfrac{92 \cdot 6}{1 \cdot 1}$

$= \dfrac{552}{1}$

$= 552$

Section 5.3

3. 20 is to 45 as 4 is to 9

$\dfrac{20}{45} \qquad \dfrac{4}{9}$

$\dfrac{20}{45} = \dfrac{4}{9}$

7. 32 is to 48 as 22 is to 33

$\dfrac{32}{48} \qquad \dfrac{22}{33}$

$\dfrac{32}{48} = \dfrac{22}{33}$

11. $\dfrac{6}{10} = \dfrac{3}{5}$

$\dfrac{6}{10} = \dfrac{2 \cdot 3}{2 \cdot 5} = \dfrac{3}{5}$

$\dfrac{3}{5} = \dfrac{3}{5}$, so the proportion is true.

15. $\dfrac{7}{10} = \dfrac{20}{30}$

$\dfrac{20}{30} = \dfrac{2 \cdot 10}{3 \cdot 10} = \dfrac{2}{3}$

$\dfrac{7}{10}$ does not equal $\dfrac{2}{3}$, so the proportion is false.

19. $\dfrac{32}{18} = \dfrac{48}{27}$

$\dfrac{32}{18} = \dfrac{2 \cdot 16}{2 \cdot 9} = \dfrac{16}{9}$

$\dfrac{48}{27} = \dfrac{3 \cdot 16}{3 \cdot 9} = \dfrac{16}{9}$

$\dfrac{16}{9} = \dfrac{16}{9}$, so the proportion is true.

23. $\dfrac{4}{7} = \dfrac{12}{21}$ cross products:

$\left. \begin{array}{l} 4 \cdot 21 = 84 \\ 7 \cdot 21 = 84 \end{array} \right]$ Same

Circle "true".

27. $\dfrac{110}{18} = \dfrac{160}{27}$ cross products:

$\left. \begin{array}{l} 110 \cdot 27 = 2970 \\ 18 \cdot 160 = 2880 \end{array} \right]$ Different

Circle "false".

31. $\dfrac{9}{16} = \dfrac{2\tfrac{7}{8}}{5}$ cross products:

$9 \cdot 5 = 45$

$\dfrac{\cancel{16}^{2}}{1} \cdot \dfrac{23}{\cancel{8}} = 46$ $\Big]$ Different

Circle "false".

35. $\dfrac{2\tfrac{5}{8}}{3\tfrac{1}{4}} = \dfrac{21}{26}$ cross products:

$\dfrac{21}{\cancel{8}} \cdot \dfrac{\cancel{26}^{13}}{1} = \dfrac{273}{4}$

$\dfrac{13}{4} \cdot \dfrac{21}{1} = \dfrac{273}{4}$ $\Big]$ Same

Circle "true".

39. $\dfrac{6.12}{24.48} = \dfrac{62.7}{250.8}$ cross products:

$(6.12)(250.8) = 1534.896$

$(24.48)(62.7) = 1534.896$ $\Big]$ Same

Circle "true".

43. $\dfrac{15}{12} = 1\dfrac{3}{12} = 1\dfrac{1}{4}$

$\begin{array}{r} 1 \ \text{R}3 \\ 12\overline{)15} \\ \underline{12} \\ 3 \end{array}$

47. $\dfrac{55}{8} = 6\dfrac{7}{8}$

$\begin{array}{r} 6 \ \text{R}7 \\ 8\overline{)55} \\ \underline{48} \\ 7 \end{array}$

Section 5.4

3. $\dfrac{k}{36} = \dfrac{7}{12}$

$12 \cdot k = 252$ *Cross products are equivalent*

$\dfrac{12 \cdot k}{12} = \dfrac{252}{12}$ *Divide both sides by 12*

$k = 21$

7. $\dfrac{10}{5} = \dfrac{x}{20}$

$\dfrac{2}{1} = \dfrac{x}{20}$ *Write 10/5 in lowest terms*

$40 = 1 \cdot x$ *Cross products are equivalent*

$40 = x$

5.4 Solving Proportions

11. $\dfrac{8}{s} = \dfrac{24}{30}$

    $\dfrac{8}{s} = \dfrac{4}{5}$ Write 24/30 in lowest terms

    $40 = 4 \cdot s$ Cross products are equivalent

    $\dfrac{40}{4} = \dfrac{4 \cdot s}{4}$ Divide both sides by 4

    $10 = s$

15. $\dfrac{y}{9} = \dfrac{62}{18}$

    $\dfrac{y}{9} = \dfrac{31}{9}$ Write 62/18 in lowest terms

    $9 \cdot y = 279$ Cross products are equivalent

    $\dfrac{9 \cdot y}{9} = \dfrac{279}{9}$ Divide both sides by 9

    $y = 31$

19. $\dfrac{1}{2} = \dfrac{k}{7}$

    $1 \cdot 7 = 2 \cdot k$ Cross products are equivalent

    $\dfrac{7}{2} = \dfrac{2 \cdot k}{2}$ Divide both sides by 2

    $3\tfrac{1}{2} = k$

23. $\dfrac{m}{3} = \dfrac{7}{8}$

    $8 \cdot m = 3 \cdot 7$ Cross products are equivalent

    $8 \cdot m = 21$

    $\dfrac{8 \cdot m}{8} = \dfrac{21}{8}$ Divide both sides by 8

    $m = 2\tfrac{5}{8}$

27. $\dfrac{10}{15} = \dfrac{y}{8}$

    $\dfrac{2}{3} = \dfrac{y}{8}$ Write 10/15 in lowest terms

    $3 \cdot y = 2 \cdot 8$ Cross products are equivalent

    $3 \cdot y = 16$

    $\dfrac{3 \cdot y}{3} = \dfrac{16}{3}$ Divide both sides by 3

    $y = 5\tfrac{1}{3}$

31. $\dfrac{2\tfrac{1}{3}}{1\tfrac{1}{2}} = \dfrac{r}{2\tfrac{1}{4}}$

    $1\tfrac{1}{2} \cdot r = 2\tfrac{1}{3} \cdot 2\tfrac{1}{4}$ Cross products are equivalent

    $\dfrac{3}{2} \cdot r = \dfrac{7}{\cancel{3}} \cdot \dfrac{\cancel{9}^{3}}{4}$

    $\dfrac{3}{2} \cdot r = \dfrac{21}{4}$

    $\dfrac{\tfrac{3}{2} \cdot r}{\tfrac{3}{2}} = \dfrac{\tfrac{21}{4}}{\tfrac{3}{2}}$ Divide both sides by 3/2

    $r = \dfrac{21}{4} \div \dfrac{3}{2}$

    $r = \dfrac{21}{4} \cdot \dfrac{2}{3}$

    $r = 3\tfrac{1}{2}$

35. $\dfrac{0}{5} = \dfrac{m}{8}$

    $0 \cdot 8 = 5 \cdot m$ Cross products are equivalent

    $0 = 5 \cdot m$

    $\dfrac{0}{5} = \dfrac{5 \cdot m}{5}$ Divide both sides by 5

    $0 = m$

Chapter 5 Ratio and Proportion

39. $\dfrac{42}{a} = \dfrac{30}{60}$

   $\dfrac{42}{a} = \dfrac{1}{2}$ *Write 30/60 in lowest terms*

   $1 \cdot a = 42 \cdot 2$ *Cross products are equivalent*

   $a = 84$

43. 10 is to 4 as 40 is to 16

   $\dfrac{10}{4}$ $\dfrac{40}{16}$

Section 5.5

3. $\dfrac{\$15}{6 \text{ magazines}} = \dfrac{x}{14 \text{ magazines}}$

   known ratio of cost per magazine   cost for 14 magazines

   $\dfrac{5}{2} = \dfrac{x}{14}$

   $2 \cdot x = 5 \cdot 14$

   $\dfrac{2 \cdot x}{2} = \dfrac{70}{2}$

   $x = 35$

   Answer: $35

7. $\dfrac{\$255.75}{5 \text{ days}} = \dfrac{x}{3 \text{ days}}$

   $\dfrac{51.15}{1} = \dfrac{x}{3}$

   $1 \cdot x = 3 \cdot 51.15$

   $x = 153.45$

   Answer: $153.45

11. $\dfrac{\$80}{4 \text{ days}} = \dfrac{x}{11 \text{ days}}$

   $4 \cdot x = 80 \cdot 11$

   $\dfrac{4 \cdot x}{4} = \dfrac{880}{4}$

   $x = 220$

   Answer: $220

15. $\dfrac{\overset{3}{\cancel{\$12}}}{\underset{20}{\cancel{80} \text{ square feet}}} = \dfrac{x}{200 \text{ square feet}}$

   $20 \cdot x = 3 \cdot 200$

   $20 \cdot x = 600$

   $\dfrac{20 \cdot x}{20} = \dfrac{600}{20}$

   $x = 30$

   Answer: $30

19. $\dfrac{\$339.30}{\$5220} = \dfrac{x}{\$7400}$

   $5220 \cdot x = 7400 \cdot 339.30$

   $\dfrac{5220 \cdot x}{5220} = \dfrac{2{,}510{,}820}{5220}$

   $x = 481$

   Answer: $481

23.
```
    2.87     2 decimal places
  × 1 00   + 0 decimal places
    0 00     2
    00 0
    287
  -------
  287.00 = 287
```

Chapter 5 Review Exercises

1. $\dfrac{6 \text{ apples}}{11 \text{ apples}} = \dfrac{6}{11}$

2. $\dfrac{12 \text{ tablets}}{8 \text{ tablets}} = \dfrac{12}{8} = \dfrac{3}{2}$

3. $\dfrac{25 \text{ miles}}{2 \text{ gallons}} = \dfrac{25}{2}$

4. $\dfrac{50 \text{ feet}}{90 \text{ feet}} = \dfrac{50}{90} = \dfrac{5}{9}$

5. $\dfrac{38 \text{ hits}}{76 \text{ times at bat}} = \dfrac{1}{2}$

6. $\dfrac{20 \text{ victories}}{30 \text{ games}} = \dfrac{2}{3}$

7. $\dfrac{5 \text{ hours}}{100 \text{ minutes}} = \dfrac{5 \cdot 60 \text{ minutes}}{100 \text{ minutes}} = \dfrac{300}{100} = 3$

8. $\dfrac{550 \text{ passengers}}{11 \text{ buses}} = 50$

9. $\dfrac{100 \text{ dollars}}{20 \text{ hours}} = 5$

10. $\dfrac{45 \text{ dollars}}{300 \text{ miles}} = \dfrac{3}{20}$

11. $\dfrac{2\tfrac{1}{2}}{5} = \dfrac{\tfrac{5}{2}}{\tfrac{5}{1}} = \dfrac{5}{2} \div \dfrac{5}{1} = \dfrac{5}{2} \cdot \dfrac{1}{5} = \dfrac{1}{2}$

12. $\dfrac{6\tfrac{1}{4}}{12\tfrac{1}{2}} = \dfrac{\tfrac{25}{4}}{\tfrac{25}{2}} = \dfrac{25}{4} \div \dfrac{25}{2} = \dfrac{25}{4} \cdot \dfrac{2}{25} = \dfrac{1}{2}$

13. $\dfrac{3 \text{ inches}}{1 \text{ foot}} = \dfrac{3 \text{ inches}}{1 \cdot 12 \text{ inches}} = \dfrac{3}{12} = \dfrac{1}{4}$

14. $\dfrac{4 \text{ feet}}{10 \text{ inches}} = \dfrac{4 \cdot 12 \text{ inches}}{10 \text{ inches}} = \dfrac{48}{10} = \dfrac{24}{5}$

15. $\dfrac{3 \text{ hours}}{30 \text{ minutes}} = \dfrac{3 \cdot 60 \text{ minutes}}{30 \text{ minutes}}$
$= \dfrac{180}{30} = 6$

16. $\dfrac{3 \text{ quarts}}{1 \text{ gallon}} = \dfrac{3 \text{ quarts}}{1 \cdot 4 \text{ quarts}} = \dfrac{3}{4}$

17. $\dfrac{20 \text{ hours}}{3 \text{ days}} = \dfrac{20 \text{ hours}}{3 \cdot 24 \text{ hours}} = \dfrac{20}{72} = \dfrac{5}{18}$

18. $\dfrac{6 \text{ quarts}}{5 \text{ pints}} = \dfrac{6 \cdot 2 \text{ pints}}{5 \text{ pints}} = \dfrac{12}{5}$

19. $\dfrac{2\tfrac{1}{2} \text{ days}}{1 \text{ week}} = \dfrac{2\tfrac{1}{2} \text{ days}}{1 \cdot 7 \text{ days}} = \dfrac{\tfrac{5}{2}}{\tfrac{7}{1}} = \dfrac{5}{2} \div \dfrac{7}{1}$
$= \dfrac{5}{2} \cdot \dfrac{1}{7} = \dfrac{5}{14}$

20. $\dfrac{3\tfrac{1}{2} \text{ yards}}{4 \text{ feet}} = \dfrac{3\tfrac{1}{2} \cdot 3 \text{ feet}}{4} = \dfrac{\tfrac{7}{2} \cdot 3}{\tfrac{4}{1}}$
$= \dfrac{21}{2} \div \dfrac{4}{1} = \dfrac{21}{2} \cdot \dfrac{1}{4} = \dfrac{21}{8}$

21. $\dfrac{\$35 \text{ in products}}{\$50 \text{ in products}} = \dfrac{35}{50} = \dfrac{7}{10}$

94   Chapter 5 Ratio and Proportion

22. $\dfrac{35 \text{ m.p.g}}{30 \text{ m.pg}} = \dfrac{35}{30} = \dfrac{7}{6}$

23. | Size | Unit Cost | |
    |------|-----------|---|
    | 16 oz. | $\dfrac{\$2.80}{16 \text{ oz.}}$ | = $.175 per ounce |
    | 8 oz. | $\dfrac{\$1.45}{8 \text{ oz.}}$ | = $.181 per ounce |
    | 3 oz. | $\dfrac{\$1.15}{3 \text{ oz.}}$ | = $3.83 per ounce |

    The best buy is the 16 ounce container.

24. | Size | Unit Cost | |
    |------|-----------|---|
    | 30 oz. | $\dfrac{\$1.50}{30 \text{ oz.}}$ | = $.05 per ounce |
    | 15 oz. | $\dfrac{\$.79}{15 \text{ oz.}}$ | = $.053 per ounce |
    | 8 oz. | $\dfrac{\$.45}{8 \text{ oz.}}$ | = $.056 per ounce |

    The best buy is the 30 ounce can.

25. 5 is to 10 as 20 is to 40.

    $\dfrac{5}{10} = \dfrac{20}{40}$

    $\dfrac{5}{10} = \dfrac{20}{40}$

26. 7 is to 2 as 35 is to 10.

    $\dfrac{7}{2} = \dfrac{35}{10}$

    $\dfrac{7}{2} = \dfrac{35}{10}$

27. $1\tfrac{1}{4}$ is to 5 as 3 is to 12.

    $\dfrac{1\tfrac{1}{4}}{5} = \dfrac{3}{12}$

    $\dfrac{1\tfrac{1}{4}}{5} = \dfrac{3}{12}$

28. $\dfrac{3}{5} = \dfrac{9}{15}$

    $\dfrac{9}{15} = \dfrac{\cancel{3} \cdot 3}{\cancel{3} \cdot 5} = \dfrac{3}{5}$

    $\dfrac{3}{5} = \dfrac{3}{5}$, so the proportion is true.

29. $\dfrac{16}{48} = \dfrac{9}{36}$

    $\dfrac{16}{48} = \dfrac{16}{16 \cdot 3} = \dfrac{1}{3}$

    $\dfrac{9}{36} = \dfrac{9}{9 \cdot 4} = \dfrac{1}{4}$

    $\dfrac{1}{3}$ does not equal $\dfrac{1}{4}$, so the proportion is false.

30. $\dfrac{47}{10} = \dfrac{98}{20}$

    $\dfrac{98}{20} = \dfrac{\cancel{2} \cdot 49}{\cancel{2} \cdot 10}$

    $\dfrac{47}{10}$ does not equal $\dfrac{49}{10}$, so the proportion is false.

31. $\dfrac{64}{36} = \dfrac{96}{54}$

    $\dfrac{64}{36} = \dfrac{4 \cdot 16}{4 \cdot 9} = \dfrac{16}{9}$

    $\dfrac{96}{54} = \dfrac{6 \cdot 16}{6 \cdot 9} = \dfrac{16}{9}$

    $\dfrac{16}{9} = \dfrac{16}{9}$, so the proportion is true.

32. $\dfrac{32}{72} = \dfrac{24}{54}$

$\dfrac{32}{72} = \dfrac{4 \cdot 8}{9 \cdot 8} = \dfrac{4}{9}$

$\dfrac{24}{54} = \dfrac{4 \cdot 6}{9 \cdot 6} = \dfrac{4}{9}$

$\dfrac{4}{9} = \dfrac{4}{9}$, so the proportion is true.

33. $\dfrac{110}{140} = \dfrac{22}{26}$

$\dfrac{110}{140} = \dfrac{11 \cdot 10}{14 \cdot 10} = \dfrac{11}{14}$

$\dfrac{22}{26} = \dfrac{2 \cdot 11}{2 \cdot 13} = \dfrac{11}{13}$

$\dfrac{11}{14}$ does not equal $\dfrac{11}{13}$, so the proportion is false.

34. $\dfrac{6}{10} = \dfrac{12}{20}$ cross products:

$6 \cdot 20 = 120$
$10 \cdot 12 = 120$ Same

The proportion is true.

35. $\dfrac{6}{8} = \dfrac{21}{28}$ cross products:

$6 \cdot 28 = 168$
$8 \cdot 21 = 168$ Same

36. $\dfrac{32}{40} = \dfrac{44}{55}$ cross products:

$40 \cdot 44 = 1760$
$32 \cdot 55 = 1760$ Same

The proportion is true.

37. $\dfrac{1}{4} = \dfrac{p}{8}$

$4 \cdot p = 8$ *Cross products are equivalent*

$\dfrac{4 \cdot p}{4} = \dfrac{8}{4}$ *Divide both sides by 4*

$p = 2$

38. $\dfrac{16}{b} = \dfrac{12}{15}$

$\dfrac{16}{b} = \dfrac{4}{5}$ *Write 12/15 in lowest terms*

$4 \cdot b = 80$ *Cross products are equivalent*

$\dfrac{4 \cdot b}{4} = \dfrac{80}{4}$ *Divide both sides by 4*

$b = 20$

39. $\dfrac{\overset{50}{\cancel{100}}}{\underset{7}{\cancel{14}}} = \dfrac{p}{56}$

$7 \cdot p = 50 \cdot 56$ *Cross products are equivalent*

$7 \cdot p = 2800$

$\dfrac{7 \cdot p}{7} = \dfrac{2800}{7}$ *Divide both sides by 7*

$p = 400$

40. $\dfrac{5}{8} = \dfrac{a}{20}$

$8 \cdot a = 5 \cdot 20$

$8 \cdot a = 100$

$\dfrac{8 \cdot a}{8} = \dfrac{100}{8}$

$a = 12\dfrac{1}{2}$

96 Chapter 5 Ratio and Proportion

41. $\dfrac{p}{24} = \dfrac{11}{18}$

   $18 \cdot p = 24 \cdot 11$

   $18 \cdot p = 264$

   $\dfrac{18 \cdot p}{18} = \dfrac{264}{18}$

   $p = 14\dfrac{2}{3}$

42. $\dfrac{f}{10} = \dfrac{2\frac{1}{2}}{8}$

   $8 \cdot f = 10 \cdot 2\dfrac{1}{2}$

   $8 \cdot f = \dfrac{10}{1} \cdot \dfrac{5}{2}$

   $8 \cdot f = 25$

   $\dfrac{8 \cdot f}{8} = \dfrac{25}{8}$

   $f = 3\dfrac{1}{8}$

43. 5 is to 8 as x is to 16.

   $\dfrac{5}{8} = \dfrac{x}{16}$

   $8 \cdot x = 5 \cdot 16$

   $8 \cdot x = 80$

   $\dfrac{8 \cdot x}{8} = \dfrac{80}{8}$

   $x = 10$

44. 7 is to 10 as 14 is to y.

   $\dfrac{7}{10} = \dfrac{14}{y}$

   $7 \cdot y = 10 \cdot 14$

   $7 \cdot y = 140$

   $\dfrac{7 \cdot y}{7} = \dfrac{140}{7}$

   $y = 20$

45. $\dfrac{16 \text{ files}}{5 \text{ minutes}} = \dfrac{x}{20 \text{ minutes}}$

   $5 \cdot x = 16 \cdot 20$

   $5 \cdot x = 320$

   $\dfrac{5 \cdot x}{5} = \dfrac{320}{5}$

   $x = 64$

   Answer: 64 files

46. $\dfrac{25 \text{ miles}}{2 \text{ hours}} = \dfrac{x}{5\frac{1}{2}}$

   $2 \cdot x = 25 \cdot 5\dfrac{1}{5}$

   $2 \cdot x = 25 \cdot \dfrac{11}{2}$

   $2 \cdot x = \dfrac{275}{2}$

   $2 \cdot x = \dfrac{\frac{275}{2}}{2}$

   $x = \dfrac{275}{2} \div \dfrac{2}{1}$

   $x = \dfrac{275}{4} \cdot \dfrac{1}{2}$

   $x = \dfrac{275}{4}$ or $68\dfrac{3}{4}$

   Answer: $68\dfrac{3}{4}$ miles

47. $\dfrac{\cancel{6}\,\overset{1}{}\text{ gallons}}{\underset{750}{\cancel{4500}}\text{ square feet}} = \dfrac{x}{13{,}500 \text{ square feet}}$

   $750 \cdot x = 1 \cdot 13{,}500$

   $\dfrac{750 \cdot x}{750} = \dfrac{13{,}500}{750}$

   $x = 18$

   Answer: 18 gallons

48. $\dfrac{\overset{19}{\cancel{\$380}} \text{ interest}}{\underset{100}{\cancel{\$2000}} \text{ loan}} = \dfrac{x}{\$3000}$

$100 \cdot x = 19 \cdot 3000$

$100 \cdot x = 57{,}000$

$\dfrac{100 \cdot x}{100} = \dfrac{57{,}000}{100}$

$x = 570$

Answer: $570 interest

49. $\dfrac{\$53}{2 \text{ hour job}} = \dfrac{x}{5 \text{ hour job}}$

$2 \cdot x = 53 \cdot 5$

$2 \cdot x = 265$

$\dfrac{2 \cdot x}{2} = \dfrac{265}{2}$

$x = 132\tfrac{1}{2}$ or 132.5

Answer: $132.50

50. $\dfrac{\$10.50}{12} = \dfrac{x}{180}$

$12 \cdot x = 10.5 \cdot 180$

$12 \cdot x = 1890$

$\dfrac{12 \cdot x}{12} = \dfrac{1890}{12}$

$x = 157\tfrac{1}{2}$ or 157.50

Answer: $157.50

51. 110 is to 140 as 22 is to 28

$\dfrac{110}{140} = \dfrac{22}{28}$

$\dfrac{110}{140} = \dfrac{22}{28}$

52. 8 is to $3\tfrac{1}{2}$ as 32 is to 14

$\dfrac{8}{3\tfrac{1}{2}} = \dfrac{32}{14}$

$\dfrac{8}{3\tfrac{1}{2}} = \dfrac{32}{14}$

53. 15 is to 5 as 21 is to 7

$\dfrac{15}{5} = \dfrac{21}{7}$

$\dfrac{15}{5} = \dfrac{21}{7}$

54. $\dfrac{p}{6} = \dfrac{8}{15}$

$15 \cdot p = 6 \cdot 8$

$15 \cdot p = 48$

$\dfrac{15 \cdot p}{15} = \dfrac{48}{15}$

$p = 3\tfrac{1}{5}$

55. $\dfrac{4}{w} = \dfrac{15}{2}$

$15 \cdot w = 4 \cdot 2$

$15 \cdot w = 8$

$\dfrac{15 \cdot w}{15} = \dfrac{8}{15}$

$w = \dfrac{8}{15}$

56. $\dfrac{b}{45} = \dfrac{\overset{7}{\cancel{70}}}{\underset{3}{\cancel{30}}}$

$3 \cdot b = 7 \cdot 45$

$3 \cdot b = 315$

$\dfrac{3 \cdot b}{3} = \dfrac{315}{3}$

$b = 105$

Chapter 5 Ratio and Proportion

57. $\dfrac{k}{20} = \dfrac{0}{20}$

$20 \cdot k = 0$ *Cross products are equal*

$\dfrac{20 \cdot k}{20} = \dfrac{0}{20}$ *Divide both sides by 20*

$k = 0$

58. $\dfrac{\overset{32}{\cancel{64}}}{\underset{5}{\cancel{10}}} = \dfrac{x}{20}$

$5 \cdot x = 32 \cdot 20$

$5 \cdot x = 640$

$\dfrac{5 \cdot x}{5} = \dfrac{640}{5}$

$x = 128$

59. $\dfrac{55}{18} = \dfrac{80}{27}$

$\left.\begin{array}{l} 18 \cdot 80 = 1440 \\ 55 \cdot 27 = 1485 \end{array}\right]$ Different

The proportion is false.

60. $\dfrac{56}{60} = \dfrac{180}{194}$

$\left.\begin{array}{l} 60 \cdot 180 = 10{,}800 \\ 56 \cdot 194 = 10{,}864 \end{array}\right]$ Different

The proportion is false.

61. $\dfrac{66}{8} = \dfrac{165}{20}$

$\left.\begin{array}{l} 8 \cdot 165 = 1320 \\ 66 \cdot 20 = 1320 \end{array}\right]$ Same

The proportion is true.

62. $\dfrac{2}{7} = \dfrac{7}{y}$

$2 \cdot y = 7 \cdot 7$

$2 \cdot y = 49$

$\dfrac{2 \cdot y}{2} = \dfrac{49}{2}$

$y = 24\dfrac{1}{2}$

63. $\dfrac{7}{1\frac{1}{2}} = \dfrac{e}{6}$

$1\dfrac{1}{2} \cdot e = 7 \cdot 6$

$\dfrac{1\frac{1}{2} \cdot e}{1\frac{1}{2}} = \dfrac{42}{1\frac{1}{2}}$

$e = 42 \div 1\dfrac{1}{2}$

$e = 42 \div \dfrac{3}{2}$

$e = \dfrac{42}{1} \cdot \dfrac{2}{3}$

$e = 28$

64. $\dfrac{3}{5} = \dfrac{z}{4\frac{1}{2}}$

$5 \cdot z = 3 \cdot 4\dfrac{1}{2}$

$5 \cdot z = 3 \cdot \dfrac{9}{2}$

$5 \cdot z = \dfrac{27}{2}$

$\dfrac{5 \cdot z}{5} = \dfrac{\frac{27}{2}}{5}$

$z = \dfrac{27}{2} \div \dfrac{5}{1}$

$z = \dfrac{27}{2} \cdot \dfrac{1}{5}$

$z = 2\dfrac{7}{10}$

65. $\dfrac{4 \text{ dollars}}{10 \text{ quarters}} = \dfrac{4 \cdot 4 \text{ quarters}}{10 \text{ quarters}}$

$= \dfrac{16}{10} = \dfrac{8}{5}$

66. $\dfrac{4\frac{1}{8}}{10} = \dfrac{\frac{33}{8}}{\frac{10}{1}} = \dfrac{33}{8} \div \dfrac{10}{1} = \dfrac{33}{8} \cdot \dfrac{1}{10} = \dfrac{33}{80}$

67. $\dfrac{10 \text{ yards}}{8 \text{ feet}} = \dfrac{10 \cdot 3 \text{ feet}}{8 \text{ feet}} = \dfrac{30}{8} = \dfrac{15}{4}$

68. $\dfrac{3 \text{ pints}}{4 \text{ quarts}} = \dfrac{3 \text{ pints}}{4 \cdot 2 \text{ pints}} = \dfrac{3}{8}$

69. $\dfrac{15 \text{ minutes}}{1 \text{ hour}} = \dfrac{15 \text{ minutes}}{1 \cdot 60 \text{ minutes}}$

$= \dfrac{15}{60} = \dfrac{1}{4}$

70. $\dfrac{1\frac{3}{10}}{4\frac{1}{2}} = \dfrac{\frac{13}{10}}{\frac{9}{2}} = \dfrac{13}{10} \div \dfrac{9}{2} = \dfrac{13}{10} \cdot \dfrac{2}{9} = \dfrac{13}{45}$

71. $\dfrac{\overset{2}{\cancel{8}} \text{ ounces medicine}}{\underset{5}{\cancel{20}} \text{ ounces water}} = \dfrac{x}{50 \text{ ounces water}}$

$5 \cdot x = 2 \cdot 50$

$5 \cdot x = 100$

$\dfrac{5 \cdot x}{5} = \dfrac{100}{5}$

$x = 20$

Answer: 20 ounces of medicine

72. $\dfrac{3 \text{ gallons}}{10 \text{ acres}} = \dfrac{x}{30 \text{ acres}}$

$10 \cdot x = 3 \cdot 30$

$10 \cdot x = 90$

$\dfrac{10 \cdot x}{10} = \dfrac{90}{10}$

$x = 9$

Answer: 9 gallons

73. $\dfrac{1 \text{ Hershey bar}}{200 \text{ calories}} = \dfrac{x}{500 \text{ calories}}$

$200 \cdot x = 1 \cdot 500$

$\dfrac{200 \cdot x}{200} = \dfrac{500}{200}$

$x = \dfrac{5}{2} \text{ or } 2\dfrac{1}{2}$

Answer: $2\dfrac{1}{2}$ Hershey bars

74. $\dfrac{\overset{10}{\cancel{40}} \text{ minutes}}{\underset{3}{\cancel{12}} \text{ car parts}} = \dfrac{x}{15 \text{ car parts}}$

$3 \cdot x = 10 \cdot 15$

$3 \cdot x = 150$

$\dfrac{3 \cdot x}{3} = \dfrac{150}{3}$

$x = 50$

Answer: 50 minutes

Chapter 5 Test

1. $\dfrac{9 \text{ ships}}{11 \text{ ships}} = \dfrac{9}{11}$

2. $\dfrac{128 \text{ feet}}{\$225} = \dfrac{128}{225}$

3. $\dfrac{12 \text{ instructions}}{8 \text{ patients}} = \dfrac{3}{2}$

4. $\dfrac{340 \text{ people}}{1120 \text{ people}} = \dfrac{17}{56}$

5. | Size | Unit Cost |
   |------|-----------|
   | 28 ounces | $\dfrac{\$2.15}{28} = \$.077$ per ounce |
   | 18 ounces | $\dfrac{\$1.67}{18} = \$.093$ per ounce |
   | 10 ounces | $\dfrac{\$1.29}{100} = \$.129$ per ounce |

   The best buy is the 28 ounce box.

6. $\dfrac{3 \text{ quarts}}{180 \text{ gallons}} = \dfrac{3 \text{ quarts}}{180 \cdot 4 \text{ gallons}}$

   $= \dfrac{3 \text{ quarts}}{720 \text{ quarts}}$

   $= \dfrac{1}{240}$

7. $\dfrac{9 \text{ inches}}{15 \text{ feet}} = \dfrac{9 \text{ inches}}{15 \cdot 12 \text{ inches}}$

   $= \dfrac{9 \text{ inches}}{180 \text{ inches}}$

   $= \dfrac{1}{20}$

8. 7 is to 15 as 21 is to 45

   $\dfrac{7}{15} = \dfrac{21}{45}$

   $\dfrac{7}{15} = \dfrac{21}{45}$

9. 12 is 7 as 36 is to 21

   $\dfrac{12}{7} = \dfrac{36}{21}$

   $\dfrac{12}{7} = \dfrac{36}{21}$

10. $\dfrac{6}{15} = \dfrac{18}{45}$

    $\dfrac{6}{15} = \dfrac{3 \cdot 2}{3 \cdot 5} = \dfrac{2}{5}$

    $\dfrac{18}{45} = \dfrac{2 \cdot 9}{5 \cdot 9} = \dfrac{2}{5}$

    Since $\dfrac{2}{5} = \dfrac{2}{5}$, the proportion is true.

11. $\dfrac{7}{20} = \dfrac{28}{75}$

    $7 \cdot 75 = 525$
    $20 \cdot 28 = 560$ $\Big\}$ Different

    The proportion is false.

12. $\dfrac{1\frac{1}{4}}{2} = \dfrac{6\frac{1}{4}}{10}$

    $1\frac{1}{4} \cdot 10 = \dfrac{5}{4} \cdot \dfrac{10}{1} = \dfrac{25}{2}$
    $2 \cdot 6\frac{1}{4} = \dfrac{2}{1} \cdot \dfrac{25}{4} = \dfrac{25}{2}$ $\Big\}$ Same

    The proportion is true.

13. $\dfrac{4\frac{3}{8}}{7\frac{1}{2}} = \dfrac{13\frac{1}{8}}{22\frac{1}{2}}$

    $4\frac{3}{8} \cdot 22\frac{1}{2} = \dfrac{35}{8} \cdot \dfrac{45}{2} = \dfrac{1575}{16}$
    $7\frac{1}{2} \cdot 13\frac{1}{8} = \dfrac{15}{2} \cdot \dfrac{105}{8} = \dfrac{1575}{16}$ $\Big\}$ Same

    The proportion is true.

14. $\dfrac{5}{9} = \dfrac{f}{45}$

$9 \cdot f = 5 \cdot 45$

$9 \cdot f = 225$

$\dfrac{9 \cdot f}{9} = \dfrac{225}{9}$

$f = 25$

15. $\dfrac{1}{3} = \dfrac{h}{8}$

$3 \cdot h = 1 \cdot 8$

$3 \cdot h = 8$

$\dfrac{3 \cdot h}{3} = \dfrac{8}{3}$

$h = 2\dfrac{2}{3}$

16. $\dfrac{6\frac{1}{2}}{4} = \dfrac{g}{20}$

$4 \cdot g = 6\dfrac{1}{2} \cdot 20$

$4 \cdot g = \dfrac{13}{2} \cdot 20$

$4 \cdot g = 130$

$\dfrac{4 \cdot g}{4} = \dfrac{130}{4}$

$g = \dfrac{65}{2}$ or $32\dfrac{1}{2}$

17. $\dfrac{\overset{48}{\cancel{240}} \text{ words}}{\underset{1}{\cancel{5} \text{ minutes}}} = \dfrac{x}{12 \text{ minutes}}$

$1 \cdot x = 48 \cdot 12$

$x = 576$

Answer: 576 words

18. $\dfrac{\overset{41}{\cancel{82}} \text{ miles}}{\underset{2}{\cancel{4} \text{ hours}}} = \dfrac{x}{9 \text{ hours}}$

$2 \cdot x = 41 \cdot 9$

$2 \cdot x = 369$

$\dfrac{2 \cdot x}{2} = \dfrac{369}{2}$

$x = \dfrac{369}{2}$ or $184\dfrac{1}{2}$ miles

19. $\dfrac{3 \text{ yards}}{2 \text{ years}} = \dfrac{x}{15 \text{ years}}$

$2 \cdot x = 3 \cdot 15$

$2 \cdot x = 45$

$\dfrac{2 \cdot x}{2} = \dfrac{45}{2}$

$x = \dfrac{45}{2}$ or 22.5 yards

20. $\dfrac{3\frac{1}{2} \text{ inches}}{7 \text{ miles}} = \dfrac{6\frac{3}{4} \text{ inches}}{x}$

$3\dfrac{1}{2} \cdot x = 6\dfrac{3}{4} \cdot 7$

$3\dfrac{1}{2} \cdot x = \dfrac{27}{4} \cdot \dfrac{7}{1}$

$3\dfrac{1}{2} \cdot x = \dfrac{189}{4}$

$\dfrac{3\frac{1}{2} \cdot x}{3\frac{1}{2}} = \dfrac{\frac{189}{4}}{3\frac{1}{2}}$

$x = \dfrac{189}{4} \div 3\dfrac{1}{2}$

$x = \dfrac{189}{4} \div \dfrac{7}{2}$

$x = \dfrac{\overset{27}{\cancel{189}}}{\underset{2}{\cancel{4}}} \cdot \dfrac{\overset{1}{\cancel{2}}}{\underset{1}{\cancel{7}}}$

$x = \dfrac{27}{2}$ or $13\dfrac{1}{2}$ miles

CHAPTER 6 PERCENT

Section 6.1

3. 25% = .25

   Drop % sign and move decimal 2 places to left.

7. 140% = 1.4

   Drop % sign and move decimal 2 places to left.

11. 14.9% = .149

    Drop % sign and move decimal 2 places to left.

15. .25% = .0025

    Two zeros are added so decimal can be moved 2 places to the left.

19. .75 = 75%

    Move decimal 2 places to the right and attach % sign.

23. .125 = 12.5%

    Move decimal 2 places to the right and attach % sign.

27. 3.82 = 382%

    Move decimal 2 places to the right and attach % sign.

31. .0312 = 3.12%

    Move decimal 2 places to the right and attach % sign.

35. .0075 = .75%

    Move decimal 2 places to the right and attach % sign.

39. 65% = .65

    Move decimal 2 places to the left and remove % sign.

43. 2.6 = 260%

    Move decimal 2 places to the right and attach % sign.

47. 47.8% = .478

    Drop % sign and move decimal 2 places to the left.

51. 3 of 10 = 30%; 7 of 10 = 70%
    (shaded)   (unshaded)

55. $2\overline{\smash{)}1.0}$ .5

    $\underline{1\ 0}$ ← zero added to dividend

    0

59. $10\overline{\smash{)}7.0}$ .7

    $\underline{7\ 0}$ ← zero added to dividend

    0

Section 6.2

3. $25\% = \frac{25}{100} = \frac{1}{4}$

7. $37.5 = \frac{37.5}{100} \cdot \frac{10}{10} = \frac{375}{1000} = \frac{3}{8}$

6.2 Percents and Fractions

11. $16\frac{2}{3}\% = \frac{16\frac{2}{3}}{100} = 16\frac{2}{3} \div 100 = \frac{50}{3} \cdot \frac{1}{100}$
$= \frac{1}{6}$

15. $.4\% = \frac{.4}{100} \cdot \frac{10}{10} = \frac{4}{1000} = \frac{1}{250}$

19. $225\% = \frac{225}{100} = \frac{9}{4}$ or $2\frac{1}{4}$

23. $\frac{7}{10} = .7 = 70\%$

27. $\frac{1}{5} = \frac{P}{100}$
$5 \cdot P = 100$ *Cross multiply*
$\frac{5 \cdot P}{5} = \frac{100}{5}$
$P = 20$
$\frac{1}{5} = 20\%$

31. $\frac{7}{8} = \frac{P}{100}$
$8 \cdot P = 7 \cdot 100$ *Cross multiply*
$8 \cdot P = 700$
$\frac{8 \cdot P}{8} = \frac{700}{8}$
$P = 87.5$
$\frac{7}{8} = 87.5\%$

35. $\frac{37}{50} = \frac{P}{100}$
$50 \cdot P = 37 \cdot 100$ *Cross multiply*
$50 \cdot P = 3700$
$\frac{50 \cdot P}{50} = \frac{3700}{50}$
$P = 74$
$\frac{37}{50} = 74\%$

39. $\frac{5}{6} = \frac{P}{100}$
$6 \cdot P = 5 \cdot 100$ *Cross multiply*
$6 \cdot P = 500$
$\frac{6 \cdot P}{6} = \frac{500}{6}$
$P = 83.33$
$\frac{5}{6} = 83.3\%$ (rounded)

43. $\frac{1}{7} = \frac{P}{100}$
$7 \cdot P = 1 \cdot 100$ *Cross multiply*
$7 \cdot P = 100$
$\frac{7 \cdot P}{7} = \frac{100}{7}$
$P = 14.28$
$\frac{1}{7} = 14.3\%$ (rounded)

47. 12.5% (given) = .125 (decimal)
$= \frac{125}{1000} = \frac{1}{8}$ (fraction)

51. $\frac{3}{10}$ (given) = .3 (decimal)
$= 30\%$ (percent)

55. 60% (given) = .6 (decimal)
$= \frac{6}{10} = \frac{3}{5}$ (fraction)

59. $\frac{3}{4}$ (given) $\quad \frac{3}{4} = \frac{P}{100}$

$\qquad 4 \cdot P = 300$

$\qquad \frac{4 \cdot P}{4} = \frac{300}{4}$

$\qquad P = 75$

$\qquad \frac{3}{4} = 75\%$ (percent)

$\qquad \quad = .75$ (decimal)

63. 90% (given) = .9 (decimal)

$\qquad = \frac{9}{10}$ (fraction)

67. .3% (given) = .003 (decimal)

$\qquad = \frac{3}{1000}$ (fraction)

71. $3\frac{1}{4}$ (given) = $\frac{13}{4}$ $\quad \frac{13}{4} = \frac{P}{100}$

$\qquad 4 \cdot P = 1300$

$\qquad \frac{4 \cdot P}{4} = \frac{1300}{4}$

$\qquad P = 325$

$\qquad \frac{13}{4} = 325\%$ (percent)

$\qquad \quad = 3.25$ (decimal)

75. $\frac{\$100}{\$500} = \frac{1}{5}$ $\quad \frac{1}{5} = \frac{P}{100}$

$\qquad 5 \cdot P = 100$

$\qquad \frac{5 \cdot P}{5} = \frac{100}{5}$

$\qquad P = 20$

$\qquad \frac{1}{5} = 20\%, \quad 20\% = .2$

The television was reduced by 1/5 or 20% or .2.

79. 40 − 22 = 18, so 18 people drive Lincolns.

$\qquad \frac{18}{40} = \frac{9}{20}$ (fraction)

$\qquad \frac{9}{20} = \frac{P}{100}$ Reduce

$\qquad 20 \cdot P = 900$ Cross multiply

$\qquad \frac{20 \cdot P}{20} = \frac{900}{20}$

$\qquad P = 45$

$\qquad \frac{9}{20} = 45\%$ (percent)

$\qquad 45\% = .45$ (decimal)

$\frac{9}{20}$ or 45% or .45 of the sales people drive Lincolns.

83. $\frac{10}{5} = \frac{X}{20}$

$\qquad \frac{2}{1} = \frac{X}{20}$ Reduce

$\qquad 40 = 1 \cdot X$ Cross multiply

$\qquad 40 = X$

87. $\frac{42}{30} = \frac{14}{b}$

$\qquad \frac{7}{5} = \frac{14}{b}$ Reduce

$\qquad 7 \cdot b = 5 \cdot 14$ Cross multiply

$\qquad 7 \cdot b = 70$

$\qquad \frac{7 \cdot b}{7} = \frac{70}{7}$

$\qquad b = 10$

6.3 The Percent Proportion

Section 6.3

3. $\dfrac{180}{b} = \dfrac{20}{100}$

    $\dfrac{180}{b} = \dfrac{1}{5}$ Reduce

    $1 \cdot b = 180 \cdot 5$

    $b = 900$

7. $\dfrac{25}{b} = \dfrac{6}{100}$

    $\dfrac{25}{b} = \dfrac{3}{50}$ Reduce

    $3 \cdot b = 25 \cdot 50$

    $3 \cdot b = 1250$

    $\dfrac{3 \cdot b}{3} = \dfrac{1250}{3}$

    $b = 416.7$ (rounded)

11. $\dfrac{105}{35} = \dfrac{p}{100}$

    $\dfrac{3}{1} = \dfrac{p}{100}$ Reduce

    $1 \cdot p = 3 \cdot 100$

    $p = 300$

15. $\dfrac{a}{52} = \dfrac{25}{100}$

    $\dfrac{a}{52} = \dfrac{1}{4}$ Reduce

    $4 \cdot a = 1 \cdot 52$

    $4 \cdot a = 52$

    $\dfrac{4 \cdot a}{4} = \dfrac{52}{4}$

    $a = 13$

19. $\dfrac{a}{47.2} = \dfrac{28}{100}$

    $\dfrac{a}{47.2} = \dfrac{7}{25}$ Reduce

    $25 \cdot a = 47.2 \cdot 7$

    $25 \cdot a = 330.4$

    $\dfrac{25 \cdot a}{25} = \dfrac{330.4}{25}$

    $a = 13.2$ (rounded)

23. $\dfrac{95}{380} = \dfrac{p}{100}$

    $\dfrac{1}{4} = \dfrac{p}{100}$ Reduce

    $4 \cdot p = 1 \cdot 100$

    $4 \cdot p = 100$

    $\dfrac{4 \cdot p}{4} = \dfrac{100}{4}$

    $p = 25$

27. $\dfrac{994.21}{8116} = \dfrac{p}{100}$

    $8116 \cdot p = 100 \cdot 994.21$

    $8116 \cdot p = 99421$

    $\dfrac{8116 \cdot p}{8116} = \dfrac{99421}{8116}$

    $p = 12.25$

31. $\dfrac{57}{100} = 57\%$

35. $\dfrac{5}{7} = \dfrac{p}{100}$

    $500 = 7p$

    $71.4 = p$ (rounded)

    $\dfrac{5}{7} = 71.4\%$ (rounded)

Section 6.4

3. 81% of what number is 748?
    - P, b, a

    P = 81
    b = unknown
    a = 748

7. 18 is 72% of what number?
    - a, P, b

    P = 72
    b = unknown
    a = 18

11. What percent of 50 is 30?
    - P, b, a

    P = unknown
    b = 50
    a = 30

15. .68% of 487 is what number?
    - P, b, a

    P = .68
    b = 487
    a = unknown

19. What percent (unknown) of $200 (price) is $8 (tax)?
    - P, b, a

23. 15% (percent of snow) of 30 inches (total snow fall) is what number (partial snow fall)?
    - P, b, a

27. $\dfrac{21}{x} = \dfrac{15}{30}$

    $\dfrac{21}{x} = \dfrac{1}{2}$ *Reduce*

    $1 \cdot x = 21 \cdot 2$

    $x = 42$

    The missing number is 42.

Section 6.5

3. 14% of 780
    = .14 · 780
    = 109.2

7. 125% of 20
    = 1.25 · 20
    = 25

11. 3% of 128
    = .03 · 128
    = 3.84

15. 15.5% of 275
    = .155 · 275
    = 42.625

6.5 Using Proportions to Solve Percent Problems

19. $\dfrac{50}{b} = \dfrac{10}{100}$

    $\dfrac{50}{b} = \dfrac{1}{10}$

    $1 \cdot b = 50 \cdot 10$

    $b = 500$

23. $\dfrac{300}{b} = \dfrac{25}{100}$

    $\dfrac{300}{b} = \dfrac{1}{4}$

    $1 \cdot b = 300 \cdot 4$

    $b = 1200$

27. $\dfrac{350}{b} = \dfrac{12\frac{1}{2}}{100}$

    $12.5 \cdot b = 350 \cdot 100$

    $12.5 \cdot b = 35{,}000$

    $\dfrac{12.5 \cdot b}{12.5} = \dfrac{35{,}000}{12.5}$

    $b = 2800$

31. $\dfrac{13}{25} = \dfrac{P}{100}$

    $25 \cdot P = 13 \cdot 100$

    $25 \cdot P = 1300$

    $\dfrac{25 \cdot P}{25} = \dfrac{1300}{25}$

    $P = 52$

    Answer: 52%

35. $\dfrac{12}{800} = \dfrac{P}{100}$

    $\dfrac{3}{200} = \dfrac{P}{100}$

    $200 \cdot P = 3 \cdot 100$

    $200 \cdot P = 300$

    $\dfrac{200 \cdot P}{200} = \dfrac{300}{200}$

    $P = \dfrac{3}{2}$ or 1.5

    Answer: 1.5%

39. $\dfrac{46}{500} = \dfrac{P}{100}$

    $500 \cdot P = 46 \cdot 100$

    $500 \cdot P = 4600$

    $\dfrac{500 \cdot P}{500} = \dfrac{4600}{500}$

    $P = 9.2$

    Answer: 9.2%

43. 38% of 2200 (total drivers) is
       P   b

    what number (number wearing seat belts)?
         a

    $\dfrac{a}{2200} = \dfrac{38}{100}$

    $100 \cdot a = 2200 \cdot 38$

    $100 \cdot a = 83{,}600$

    $\dfrac{100 \cdot a}{100} = \dfrac{83{,}600}{100}$

    $a = 836$ drivers

47. 110% of what number (applicants
     P   b

    last year) is 550 (applicants
               a

    this year)?

$$\frac{550}{b} = \frac{110}{100}$$

$$\frac{550}{b} = \frac{11}{10}$$

$$11 \cdot b = 550 \cdot 10$$

$$11 \cdot b = 5500$$

$$\frac{11 \cdot b}{11} = \frac{5500}{11}$$

b = 500 applicants last year

51. $\frac{11,700}{755,000} = \frac{p}{100}$

p = 1.5%   (rounded)

55. 72% of what number (total students)
       P   b

is 108 (part that passed)?
   a

$$\frac{108}{b} = \frac{72}{100}$$

$$72 \cdot b = 108 \cdot 100$$

$$72 \cdot b = 10,800$$

$$\frac{72 \cdot b}{72} = \frac{10,800}{72}$$

b = 150 students in the class

59. 100% total shipment
    − 35% damaged posters
       65% undamaged posters

65% of 175
= .65 · 175
= 113.75

$113.75 was the value of the undamaged posters.

63.   11
    325.6   1 decimal place
  × .031   + 3 decimal places
    3256   4 in answer
    9 768
   10.0936

67.   .25
  344)86.00 Add two zeros
    68 8   to dividend
    17 20
    17 20
       0

Section 6.6

3. 85% of 900

   .85 · 900 = a
         765 = a

7. 8% of 140

   .08 · 140 = a
         11.2 = a

11. 12.4% of 8300

    .124 · 8300 = a
         1029.2 = a

15. .3% of 480

    .003 · 480 = a
         1.44 = a

6.7 Applications of Percent

19. 75% of what number is 375.

$.75 \cdot b = 375$

$\dfrac{.75 \cdot b}{.75} = \dfrac{375}{.75}$

$b = 500$

23. $6\tfrac{1}{4}$% of what number is 25.

$.0625 \cdot b = 25$

$\dfrac{.0625 \cdot b}{.0625} = \dfrac{25}{.0625}$

$b = 400$

27. 30 is what percent of 75.

$30 = P \cdot 75$

$\dfrac{30}{75} = \dfrac{P \cdot 75}{75}$

$.4 = P$

Answer: 40%

31. What percent of 250 is 87.5.

$P \cdot 250 = 87.5$

$\dfrac{P \cdot 250}{250} = \dfrac{87.5}{250}$

$P = .35$

Answer: 35%

35. 1225 is what percent of 850.

$1224 = P \cdot 850$

$\dfrac{1224}{850} = \dfrac{P \cdot 850}{850}$

$1.44 = P$

Answer: 144%

39. $135 = p \cdot 500$

$.27 = p$

$27\% = p$

43. Find the increase in the mileage, then add it to the old mileage.

$\dfrac{15}{100} = \dfrac{a}{25.6}$

$100 \cdot a = 15 \cdot 25.6$

$100 \cdot a = 384$

$\dfrac{100 \cdot a}{100} = \dfrac{384}{100}$

$a = 3.84 = 3.8$ (rounded)

Then, $25.6 + 3.8 = 29.4$.

Mileage of 29.4 mpg can be expected.

47. 36 is 72% of what number.

$a \quad P \quad b$

$P = 72$

$b =$ unknown

$a = 36$

Section 6.7

3. Tax = $50 \cdot 5\%$
= $50 \cdot .05$
= $2.50

Total cost = $50 + 2.50 = \$52.50$

7. Tax = $10 \cdot 4\%$
= $10 \cdot .04$
= $.40

Total cost = $10 + .40 = \$10.40$

11. $1000 \cdot 22\% = 1000 \cdot .22$
= $220 commission

Chapter 6 Percent

15. $6225 \cdot 10\% = 6225 \cdot .10$
$= \$622.50$ commission

19. Discount $= \$780 \cdot 35\%$
$= 780 \cdot .35$
$= \$273$

Amount $= 780 - 273 = \$507$

23. Discount $= \$12.50 \cdot 10\%$
$= 12.50 \cdot .10$
$= \$1.25$

Amount paid $= 12.50 - 1.25 = \$11.25$

27. $45 \cdot 30\% = 45 \cdot .30$
$= \$13.50$ discount
$45 - 13.50 = \$31.50$ price after discount

31. $3286 - 2480 = 806$, so 806 is the increase in the number of students. Now convert to a percent, 806 as compared to 2480 (the previous enrollment).

$$\frac{806}{2480} = \frac{P}{100}$$
$2480 \cdot P = 806 \cdot 100$
$2480 \cdot P = 80,600$
$$\frac{2480 \cdot P}{2480} = \frac{80,600}{2480}$$
$P = 32.5$

Enrollment increased by 32.5%.

35. Rate of commission $\cdot \$4000 = \$240$
      P   b   a

$$\frac{240}{4000} = \frac{P}{100}$$
$$\frac{24}{400} = \frac{P}{100}$$
$400 \cdot P = 24 \cdot 100$
$400 \cdot P = 2400$
$$\frac{400 \cdot P}{400} = \frac{2400}{400}$$

The rate of commission is 6%.

39. $41.1 - 40.9 = .2$, which is the amount of decrease. Compare this to 40.9 to find the percent of decrease.

$$\frac{.2}{40.9} = \frac{P}{100}$$
$40.9 \cdot P = .2 \cdot 100$
$40.9 \cdot P = 20$
$$\frac{40.9 \cdot P}{40.9} = \frac{20}{40.9}$$
$P = .5$ (rounded)

.5% is the percent of decrease.

43. Discount $= 35\%$ of $\$960$
$= .35 \cdot 960$
$= \$336$

Sales price $= 960 - 336 = \$624$.

47. Member fee is 6% of selling price
      a   is 6% of   8,680,000

$$\frac{a}{8,680,000} = \frac{6}{100}$$
$100 \cdot a = 52,080,000$
$$\frac{100 \cdot a}{100} = \frac{52,080,000}{100}$$
$a = 520,800$

Association fee is 2% of member fee
         | | |
         a   is 2% of 520,800

$$\frac{a}{520,800} = \frac{2}{100}$$

$$100 \cdot a = 1,041,600$$

$$\frac{100 \cdot a}{100} = \frac{1,041,600}{100}$$

$$a = 10,416$$

The association will get $10,416.

51. .3% of 960
         |
     .003 · 960 = a
         2.88 = a

55. 147.2 is what percent of 460.
         | | |
       147.2   =   P   · 460

$$\frac{147.2}{460} = \frac{P \cdot 460}{460}$$

$$.32 = P$$

$$P = 32\%$$

Section 6.8

3. I = P · r · t
       | | |
    = $500 · 12% · 4
    = 500 · .12 · 4
    I = $240

7. I = P · r · t
    = $1500 · 13% · 6
    = 1500 · .13 · 6
    I = $1170

11. I = P · r · t
       | | |
    = $620 · 15% · $1\frac{1}{4}$
    = 620 · .15 · 1.25
    I = $116.25

15. I = P · r · t
    = $500 · 11% · $\frac{12}{12}$
    = 500 · .11 · 1
    I = $55

19. I = P · r · t
       | | |
    = $780 · 16% · $\frac{10}{12}$
    = 780 · .16 · $\frac{5}{6}$
    I = $104

23. I = P · r · t
    = $14,400 · 8% · $\frac{7}{12}$
    = 14,500 · .08 · $\frac{7}{12}$
    I = $676.67   (rounded)

27. First find interest.
    I = $780 · 12% · $\frac{3}{12}$
      = 780 · .12 · $\frac{1}{4}$
      = $23.40
    Second find amount due.
    Amount due = principal + interest
               = 780 + 23.40
               = $803.40

31. 1st: Interest = $2210 · 7% · $\frac{6}{12}$

    = 2210 · .07 · $\frac{1}{2}$

    = $77.35

    2nd: Add interest to principal.
    77.35 + 2210 = $2287.35

35. 1st: Interest = $18,200 · 16% · $\frac{8}{12}$

    = 18,200 · .16 · $\frac{2}{3}$

    = $1941.33   (rounded)

    2nd: Add interest to principal.
    1941.33 + 18,200
    = $20,141.33

39. Interest = $6500 · 16% · 1.5
    (18 months = 1.5 years)
    = 6500 · .16 · 1.5
    = $1560

43. Interest = $7500 · 9% · $\frac{6}{12}$

    = 7500 · .09 · $\frac{1}{2}$

    = $337.50

47. Interest = $4500 · 11.25% · $\frac{5}{12}$

    = 4500 · .1125 · $\frac{5}{12}$

    = $210.94

51. $\frac{5}{12} = \frac{?}{48}$ 48 ÷ 12 = 4

    $\frac{5 \cdot 4}{12 \cdot 4} = \frac{20}{48}$

55. $\frac{15}{19} = \frac{?}{76}$ 76 ÷ 19 = 4

    $\frac{15 \cdot 4}{19 \cdot 4} = \frac{60}{76}$

Chapter 6   Review Exercises

1. 50% = .5
   Drop % sign and move decimal 2 places to left.

2. 250% = 2.5
   Drop % sign and move decimal 2 places to left.

3. 17.6% = .176
   Drop % sign and move decimal 2 places to left.

4. .036% = .00036
   Two zeros are added so decimal can be moved 2 places to the left.

5. 2.8 = 280%
   Move decimal 2 places to the right and attach % sign.

6. .09 = 9%
   Move decimal 2 places to the right and attach % sign.

7. .375 = 37.5%
   Move decimal 2 places to the right and attach % sign.

8. .002 = .2%
   Move decimal 2 places to the right and attach % sign.

9. $38\% = \frac{38}{100} = \frac{19}{50}$

10. $6.25\% = \frac{6.25}{100} \cdot \frac{100}{100} = \frac{625}{10,000} = \frac{1}{16}$

11. $66\frac{2}{3}\% = \frac{66\frac{2}{3}}{100} = 66\frac{2}{3} \div 100$

    $= \frac{200}{3} \cdot \frac{1}{100} = \frac{2}{3}$

12. $.025\% = \frac{.025}{100} = \frac{25}{100,000} = \frac{1}{4000}$

13. $\frac{3}{4} = \frac{P}{100}$

    $4 \cdot P = 300$ *Cross multiply*

    $\frac{4 \cdot P}{4} = \frac{300}{4}$

    $P = 75$

    $\frac{3}{4} = 75\%$

14. $\frac{7}{8} = \frac{P}{100}$

    $8 \cdot P = 7 \cdot 100$ *Cross multiply*

    $8 \cdot P = 700$

    $\frac{8 \cdot P}{8} = \frac{700}{8}$

    $P = 87.5$

    $\frac{7}{8} = 87.5\%$

15. $\frac{7}{10} = \frac{P}{100}$

    $10 \cdot P = 7 \cdot 100$

    $10 \cdot P = 700$

    $\frac{10 \cdot P}{10} = \frac{700}{10}$

    $P = 70$

    $\frac{7}{10} = 70\%$

16. $\frac{1}{200} = \frac{P}{100}$

    $200 \cdot P = 1 \cdot 100$

    $200 \cdot P = 100$

    $\frac{200 \cdot P}{200} = \frac{100}{200}$

    $P = .5$

    $\frac{1}{200} = .5\%$

17. $\frac{1}{8}$ $\quad 8\overline{)1.000}^{\;.125}$ $\quad \frac{1}{8} = .125$

    $\phantom{\frac{1}{8} \quad 8\overline{)1.000}}\underline{8}$
    $\phantom{\frac{1}{8} \quad 8\overline{)1.000\;}}20$
    $\phantom{\frac{1}{8} \quad 8\overline{)1.000\;}}\underline{16}$
    $\phantom{\frac{1}{8} \quad 8\overline{)1.000\;\,}}40$
    $\phantom{\frac{1}{8} \quad 8\overline{)1.000\;\,}}\underline{40}$
    $\phantom{\frac{1}{8} \quad 8\overline{)1.000\;\,\,}}0$

18. $\frac{1}{8} = \frac{P}{100}$ $\quad \frac{1}{8} = 12.5\%$

    $8 \cdot P = 100$

    $\frac{8 \cdot P}{8} = \frac{100}{8}$

    $P = 12.5$

19. $.05 = \frac{5}{100} = \frac{1}{20}$

20. $.05 = 5\%$

21. $87.5\% = \frac{87.5}{100} \cdot \frac{10}{10} = \frac{875}{1000} = \frac{7}{8}$

22. $87.5\% = .875$

23. $\frac{100}{b} = \frac{10}{100}$

    $\frac{100}{b} = \frac{1}{10}$

    $1 \cdot b = 100 \cdot 10$

    $b = 1000$

114   Chapter 6   Percent

24. $\dfrac{a}{480} = \dfrac{20}{100}$

$\dfrac{a}{480} = \dfrac{1}{5}$

$5 \cdot a = 480 \cdot 1$

$5 \cdot a = 480$

$\dfrac{5 \cdot a}{5} = \dfrac{480}{5}$

$a = 96$

25. $\underset{P}{40\%}$ of $\underset{b}{150}$ is $\underset{a}{60}$.

P = 40
b = 150
a = 60

26. $\underset{a}{73}$ is $\underset{P}{\text{what percent}}$ of $\underset{b}{90}$?

P = unknown
b = 90
a = 73

27. Find $\underset{P}{46\%}$ of $\underset{b}{1040}$.

P = 46
b = 1040
a = unknown

28. $\underset{a}{418}$ is $\underset{P}{30\%}$ of $\underset{b}{\text{what number}}$?

P = 30
b = unknown
a = 418

29. A golfer lost $\underset{a}{3}$ of his $\underset{b}{8}$ balls.

$\underset{P}{\text{What percent}}$ were lost?

P = unknown
b = 8
a = 3

30. $\underset{P}{78\%}$ of $\underset{b}{640}$ will bloom. $\underset{a}{\text{How many}}$ will bloom?

P = 78
b = 640
a = unknown

31. 10% of 600
    = .10 · 600
    = 60

32. 92% of 3080
    = .92 · 3080
    = 2833.6

33. .9% of 4800
    = .009 · 4800
    = 43.2

34. .2% of 1400
    = .002 · 1400
    = 2.8

35. $\dfrac{50}{b} = \dfrac{10}{100}$

$\dfrac{50}{b} = \dfrac{1}{10}$

$1 \cdot b = 50 \cdot 10$

$b = 500$

36. $\dfrac{128}{b} = \dfrac{5}{100}$

$\dfrac{128}{b} = \dfrac{1}{20}$

$1 \cdot b = 128 \cdot 20$

$b = 2560$

Chapter 6 Review Exercises 115

37. $\dfrac{338.8}{b} = \dfrac{140}{100}$

$\dfrac{338.8}{b} = \dfrac{7}{5}$

$7 \cdot b = 338.8 \cdot 5$

$7 \cdot b = 1694$

$\dfrac{7 \cdot b}{7} = \dfrac{1694}{7}$

$b = 242$

38. $\dfrac{425}{b} = \dfrac{2.5}{100}$

$\dfrac{425}{b} = \dfrac{25}{1000}$

$\dfrac{425}{b} = \dfrac{1}{40}$

$1 \cdot b = 425 \cdot 40$

$b = 17{,}000$

39. $\dfrac{75}{150} = \dfrac{P}{100}$

$\dfrac{1}{2} = \dfrac{P}{100}$

$2 \cdot P = 1 \cdot 100$

$2 \cdot P = 100$

$\dfrac{2 \cdot P}{2} = \dfrac{100}{2}$

$P = 50$

Answer: 50%

40. $\dfrac{30}{2400} = \dfrac{P}{100}$

$\dfrac{1}{80} = \dfrac{P}{100}$

$80 \cdot P = 1 \cdot 100$

$80 \cdot P = 100$

$\dfrac{80 \cdot P}{80} = \dfrac{100}{80}$

$P = 1.25$

Answer: 1.3% (rounded)

41. $\dfrac{18}{190} = \dfrac{P}{100}$

$190 \cdot P = 18 \cdot 100$

$190 \cdot P = 1800$

$\dfrac{190 \cdot P}{190} = \dfrac{1800}{190}$

$P = 9.47$

Answer: 9.5% (rounded)

42. $\dfrac{150}{720} = \dfrac{P}{100}$

$\dfrac{5}{24} = \dfrac{P}{100}$

$24 \cdot P = 5 \cdot 100$

$24 \cdot P = 500$

$\dfrac{24 \cdot P}{24} = \dfrac{500}{24}$

$P = 20.83$

Answer: 20.8% (rounded)

43. $\underset{P}{\underline{27\%}}$ of $\underset{b}{\underline{\$32{,}000}}$ (total earnings)

is $\underset{a}{\underline{\text{what number}}}$ (tax)?

$\dfrac{a}{32{,}000} = \dfrac{27}{100}$

$100 \cdot a = 32{,}000 \cdot 27$

$100 \cdot a = 864{,}000$

$\dfrac{100 \cdot a}{100} = \dfrac{864{,}000}{100}$

$a = \$8640$ tax

44. Interest = $\$1230 \cdot 6.3\% \cdot 1$

= $1230 \cdot .063 \cdot 1$

= $\$77.49$

45. 46% of 84

$.46 \cdot 84 = a$

$38.64 = a$

46. 114% of 32

$1.14 \cdot 32 = a$

$36.48 = a$

47. .128 is what percent of 32?

$.128 = P \cdot 32$

$\dfrac{.128}{32} = \dfrac{P \cdot 32}{32}$

$.004 = P$

$P = .4\%$

48. 75 is what percent of 60?

$75 = P \cdot 60$

$\dfrac{75}{60} = \dfrac{P \cdot 60}{60}$

$\dfrac{5}{4} = P$

$\dfrac{5}{4} = 1.25$

$P = 125\%$

49. 33.6 is 28% of what number?

$33.6 = .28 \cdot b$

$\dfrac{33.6}{.28} = \dfrac{.28 \cdot b}{.28}$

$120 = b$

50. 46 is 32% of what number?

$46 = .32 \cdot b$

$46 = .32 \cdot b$

$\dfrac{46}{.32} = \dfrac{.32 \cdot b}{.32}$

$143.75 = b$

51. Tax = $100 \cdot 3\%$

   = $100 \cdot .03$

   = $3

Total cost = 100 + 3 = $103

52. Tax = $57 \cdot 2\%$

   = $57 \cdot .02$

   = $1.14

Total cost = 57 + 1.14 = $58.14

53. Commission = $360 \cdot 25\%$

   = $360 \cdot .25$

   = $90

54. Commission = $18,950 \cdot 6\%$

   = $18,950 \cdot .06$

   = $1137

55. Discount = $100 \cdot 10\%$

   = $100 \cdot .1$

   = $10

Amount paid = 100 − 10 = $90

56. Discount = $585.50 \cdot 15\%$

   = $585.50 \cdot .15$

   = $87.83 (rounded)

Amount paid = 585.50 − 87.83

   = $497.67

57. $I = p \cdot r \cdot t$

   = $100 \cdot 14\% \cdot 1$

   = $100 \cdot .14 \cdot 1$

   = $14

58. $I = \$2800 \cdot 18\% \cdot 4\frac{1}{2}$
    $= 2800 \cdot .18 \cdot 4.5$
    $= \$2268$

59. $I = \$100 \cdot 15\% \cdot \frac{6}{12}$
    $= 100 \cdot .15 \cdot .5$
    $= \$7.50$

60. $I = \$1150 \cdot 10\% \cdot \frac{18}{12}$
    $= 1150 \cdot .10 \cdot 1.5$
    $= \$172.50$

61. 1st: Interest $= \$200 \cdot 17\% \cdot 2$
    $= 200 \cdot .17 \cdot 2$
    $= \$68$

    2nd: Add interest to principal.
    $68 + 200 = \$268$
    Total amount due $= \$268$

62. 1st: Interest $= \$1420 \cdot 18\% \cdot \frac{2}{12}$
    $= 1420 \cdot .18 \cdot \frac{1}{6}$
    $= \$42.60$

    2nd: Add interest to principal.
    $42.60 + 1420 = \$1462.60$

63. $\frac{a}{50} = \frac{70}{100}$
    $\frac{a}{50} = \frac{7}{10}$
    $10 \cdot a = 50 \cdot 7$
    $10 \cdot a = 350$
    $\frac{10 \cdot a}{10} = \frac{350}{10}$
    $a = 35$

64. $\frac{350}{b} = \frac{25}{100}$
    $\frac{350}{b} = \frac{1}{4}$
    $1 \cdot b = 350 \cdot 4$
    $b = 1400$

65. 24% of 97
    $.24 \cdot 97 = a$
    $23.28 = a$

66. 195 is what percent of 130?
    $195 = P \cdot 130$
    $\frac{195}{130} = \frac{P \cdot 130}{130}$
    $1.5 = P$
    $P = 150\%$

67. .6% of 85
    $.006 \cdot 85 = a$
    $.51 = a$

68. 125 is 40% of what number?
    $125 = .40 \cdot b$
    $125 = .40 \cdot b$
    $\frac{125}{.40} = \frac{.40 \cdot b}{.40}$
    $312.5 = b$

69. 38 is what percent of 95?

   $38 = P \cdot 95$

   $38 = P \cdot 95$

   $\dfrac{38}{95} = \dfrac{P \cdot 95}{95}$

   $.4 = P$

   Answer: 40%

70. 107.242 is 43% of what number?

   $107.242 = .43 \cdot b$

   $\dfrac{107.242}{.43} = \dfrac{.43 \cdot b}{.43}$

   $249.4 = b$

71. $25\% = .25$

   Drop % sign and move decimal 2 places to left.

72. $100\% = 1.00$ or 1

   Drop % sign and move decimal 2 places to left.

73. $.50 = 50\%$

   Move decimal 2 places to the right and attach % sign.

74. $6.8 = 680\%$

   Move decimal 2 places to the right and attach % sign.

75. $8.5\% = .085$

   One zero is added so decimal can be moved 2 places to the left. Drop % sign.

76. $.719 = 71.9\%$

   Move decimal 2 places to the right and attach % sign.

77. $.25\% = .0025$

   Two zeros are added so decimal can be moved 2 places to the left. Drop % sign.

78. $.0006 = .06\%$

   Move decimal 2 places to the right and attach % sign.

79. $45\% = \dfrac{45}{100} = \dfrac{9}{20}$

80. $\dfrac{1}{2} = \dfrac{P}{100}$

   $2 \cdot P = 1 \cdot 100$ Cross multiply

   $2 \cdot P = 100$

   $\dfrac{2 \cdot P}{2} = \dfrac{100}{2}$

   $P = 50$

   $\dfrac{1}{2} = 50\%$

81. $37.5\% = \dfrac{37.5}{100} \cdot \dfrac{10}{10} = \dfrac{375}{1000} = \dfrac{3}{8}$

82. $\dfrac{3}{8} = \dfrac{P}{100}$

   $8 \cdot P = 3 \cdot 100$ Cross multiply

   $8 \cdot P = 300$

   $\dfrac{8 \cdot P}{8} = \dfrac{300}{8}$

   $P = 37.5$

   $\dfrac{3}{8} = 37.5\%$

83. $14\frac{1}{2}\% = \dfrac{14\frac{1}{2}}{100} = 14\frac{1}{2} \div 100$

$= \dfrac{29}{2} \cdot \dfrac{1}{100} = \dfrac{29}{200}$

84. $\dfrac{1}{5} = \dfrac{P}{100}$

$5 \cdot P = 1 \cdot 100$ *Cross multiply*

$5 \cdot P = 100$

$\dfrac{5 \cdot P}{5} = \dfrac{100}{5}$

$P = 20$

$\dfrac{1}{5} = 20\%$

85. $.5\% = \dfrac{.5}{100} \cdot \dfrac{10}{10} = \dfrac{5}{1000} = \dfrac{1}{200}$

86. $\dfrac{1}{400} = \dfrac{P}{100}$

$400 \cdot P = 1 \cdot 100$

$400 \cdot P = 100$

$\dfrac{400 \cdot P}{400} = \dfrac{100}{400}$

$P = .25$

$\dfrac{1}{400} = .25\% = .3\%$ (rounded)

87. $I = \$5000 \cdot 13\% \cdot \dfrac{18}{12}$

$= 5000 \cdot .13 \cdot 1.5$

$= \$975$

88. 1st: Interest $= \$1000 \cdot 15\% \cdot \dfrac{3}{12}$

$= 1000 \cdot .15 \cdot \dfrac{1}{4}$

$= \$37.50$

2nd: Add interest to amount borrowed.

$37.50 + 1000 = \$1037.50$

89. Original amount $= 38 + 342 = \$380$

What percent of $\$380$ (original amount) is $\$342$ (amount remaining)?

$\dfrac{342}{380} = \dfrac{P}{100}$

$380 \cdot P = 342 \cdot 100$

$380 \cdot P = 34,200$

$\dfrac{380 \cdot P}{380} = \dfrac{34,200}{380}$

$P = 90$

Answer: 90% still remains.

90. 3.28 billion − 3.12 billion = .16 billion, so .16 billion is the amount of increase; compare this to the original amount.

$\dfrac{.16 \text{ billion}}{3.12 \text{ billion}} = \dfrac{P}{100}$

$\dfrac{.16}{3.12} = \dfrac{P}{100}$

$3.12 \cdot P = .16 \cdot 100$

$3.12 \cdot P = 16$

$\dfrac{3.12 \cdot P}{3.12} = \dfrac{16}{3.12}$

$P = 5.1$ (rounded)

Answer: 5.1%

91. Total sales are 105,000 + 145,000 = 250,000.

Commission is $1\frac{1}{2}$% of total sales.

a = .015 · 250,000

a = 3750

$3750 is the commission.

92. 472 − 436 = 36, so 36 stops is the amount of the decrease; compare this to the original 472 stops.

$$\frac{36}{472} = \frac{P}{100}$$

$472 \cdot P = 36 \cdot 100$

$$\frac{472 \cdot P}{472} = \frac{3600}{472}$$

$P = 7.6$ (rounded)

Answer: 7.6%

93. 34.9 − 34.7 = .2, so .2 hours is the amount of increase; compare this to the original 34.7 hours.

$$\frac{.2}{34.7} = \frac{P}{100}$$

$34.7 \cdot P = .2 \cdot 100$

$$\frac{34.7 \cdot P}{34.7} = \frac{20}{34.7}$$

$P = .6$ (rounded)

Answer: .6%

94. Sales tax is 6% of total sales.

478.20 = .06 · b

$$\frac{478.20}{.06} = \frac{.06 \cdot b}{.06}$$

7970 = b

$7970 is the total sales.

95. 25% + 30% + 8% + 20% = 83% and 100% − 83% = 17%, so the couple saves 17% of their earnings.

$950 · 12 + $14,500
  = $11,400 + $14,500
  = $25,900,

so $25,900 is the couple's yearly income.

They save 17% of 25,900
or   .17 · 25,900 = $4403.

96. 16.33 billion − 15.22 billion = 1.11 billion, so 1.11 billion miles is the amount of increase; compare this to the original 15.22 billion miles.

$$\frac{1.11}{15.22} = \frac{P}{100}$$

$15.22 \cdot P = 100 \cdot 1.11$

$$\frac{15.22 \cdot P}{15.22} = \frac{111}{15.22}$$

$P = 7.3$ (rounded)

Answer: 7.3%

Chapter 6 Test

1. $75\% = .75$

2. $.05\% = .0005$

3. $.25 = 25\%$

4. $.375 = 37.5\%$

5. $37.5\% = \frac{37.5}{100} \cdot \frac{10}{10} = \frac{375}{1000} = \frac{3}{8}$

6. $.5\% = \frac{.5}{100} \cdot \frac{10}{10} \cdot \frac{5}{1000} = \frac{1}{200}$

7. $\frac{1}{2} = \frac{P}{100}$

   $2 \cdot P = 1 \cdot 100$

   $\frac{2 \cdot P}{2} = \frac{100}{2}$

   $P = 50$

   Answer: 50%

8. $\frac{5}{8} = \frac{P}{100}$

   $8 \cdot P = 5 \cdot 100$

   $\frac{8 \cdot P}{8} = \frac{500}{8}$

   $P = 62.5$

   Answer: 62.5%

9. $\frac{a}{b} = \frac{P}{100}$

   $\frac{81}{b} = \frac{30}{100}$

   $30 \cdot b = 81 \cdot 100$

   $\frac{30 \cdot b}{30} = \frac{8100}{30}$

   $b = 270$ (rounded)

10. $\frac{a}{550} = \frac{25}{100}$

    $\frac{a}{550} = \frac{1}{4}$

    $\frac{4 \cdot a}{4} = \frac{550}{4}$

    $a = 137.5$

11. $\underset{P}{15\%}$ of $\underset{b}{920}$ is $\underset{a}{138}$.

    $P = 15$

    $b = 920$

    $a = 138$

12. 8% of $200

    $= .08 \cdot 200$

    $= 16$

13. $\underset{a}{75}$ is $\underset{P}{10\%}$ of $\underset{b}{\text{what number?}}$

    $\frac{75}{b} = \frac{10}{100}$

    $\frac{75}{b} = \frac{1}{10}$

    $1 \cdot b = 75 \cdot 10$

    $b = 750$

14. $\underset{a}{150}$ is $\underset{P}{\text{what percent}}$ of $\underset{b}{600?}$

    $\frac{150}{600} = \frac{P}{100}$

    $\frac{1}{4} = \frac{P}{100}$

    $4 \cdot P = 1 \cdot 100$

    $\frac{4 \cdot P}{4} = \frac{100}{4}$

    $P = 25$

    Answer: 25%

15. 72% of total is 12,096.

    .72 • b = 12,096

    $\dfrac{.72 \cdot b}{.72} = \dfrac{12{,}096}{.72}$

    b = 16,800

    The total down payment is $16,800.

16. Tax = $28 • 4%
    = 28 • .04
    = $1.12

    Total cost = 28 + 1.12 = $29.12

17. Tax = $73 • 5%
    = 73 • .05
    = $3.65

    Total cost = 73 + 3.65 = $76.65

18. 112.50 is what percent of 1875?

    112.50 = P • 1875

    $\dfrac{112.50}{1875} = \dfrac{P \cdot 1875}{1875}$

    .06 = P

    .06 = 6%

    The rate of sales tax is 6%.

19. Commission = $2460 • 16%
    = 2460 • .16
    = $393.60

20. 1944 − 1440 = 504, so the increase is 504 students; compare this to the original 1440 students.

    $\dfrac{504}{1440} = \dfrac{P}{100}$

    1440 • P = 504 • 100

    $\dfrac{1440 \cdot P}{1440} = \dfrac{50{,}400}{1440}$

    P = 35

    Answer: 35%

21. Discount = $48 • 8%
    = 48 • .08
    = $3.84

    Amount paid = 48 − 3.84 = $44.16

22. Discount = $175 • 37.5%
    = 175 • .375
    = $65.63

    Amount paid = 175 − 65.63 = $109.37

23. I = $480 • 15% • $2\tfrac{1}{2}$
    = 480 • .15 • 2.5
    = $180

24. I = $1750 • 16% • $\tfrac{9}{12}$
    = 1750 • .16 • $\tfrac{3}{4}$
    = $210

25. Interest = $2800 • 14% • $\tfrac{6}{12}$
    = 2800 • .14 • .5
    = $196

    $196 interest will be owed.

26. Interest = $17,200 · 11% · 1
    = 17,200 · .11 · 1
    = $1892

   Total due = 17,200 + 1892 = $19,092

Cumulative Review Chapters 4–6

1. 8.64
   tenths: 6 (first digit to right of decimal point)
   hundredths: 4 (second digit to right of decimal point)

2. 93.4718
   tenths: 4 (first digit to right of decimal point)
   ten-thousandths: 8 (fourth digit to right of decimal point)

3. $.75 = \frac{75}{100} = \frac{3}{4}$

4. $.125 = \frac{125}{1000} = \frac{1}{8}$

5. $.04 = \frac{4}{100} = \frac{1}{25}$

6. $.875 = \frac{875}{1000} = \frac{7}{8}$

7. 61.6$\underline{2}$8 four or less, 6 does not change.
   Drop digits to the right of six.
   Answer: 61.6

8. .659$\underline{6}$ 5 or greater, 9 becomes 10 which means 59 becomes 60. Drop the digit to the right of nine.
   Answer: .660

9. $25.$\underline{4}$9 4 or less, 5 does not change.
   Answer: $25

10. $182.$\underline{5}$4 5 or greater, 2 becomes 3.
    Answer: $183

11. $4729.$\underline{8}$7 5 or greater, 9 becomes 10 which means 29 becomes 30.
    Answer: $4730

12.   17.63
       8.79
    + 64.52
      90.94

13.   9.030
       38.620   Attach 0's as
      835.900   necessary
        3.609
    + 73.170
      960.329

14.   32.60
      15.92   Line up decimals, attach
       3.70   0's as needed
    + 6.04
      58.26

15.   2 11
        26.3̸1̸
      - 15.1 7
        11.1 4

124   Chapter 6   Percent

```
           12
         4̸2̸ 14 5 12
16.     9 5̸ 8̸. 4̸ 6̸ 2̸
      - 1 4 8. 6 3 8
        8 0 4. 8 2 4
```

```
          7 11
17.     13̸8̸. 1̸6      Line up
       -  75. 52      decimals
           62. 64
```

```
18.      .836         3 decimal places
       × 3.2          + 1 decimal place
         1672         4 in answer
         2508
        2.6752
```

```
19.      72.71        2 decimal places
       ×  .305        + 3 decimal places
         36355        5 in answer
        218130
        22.17655
```

```
20.      .079         3 decimal places
       ×  .006        + 3 decimal places
         .000474      6 in answer
       3 zeros must be attached.
```

```
            18.36     Bring decimal point
21.     7)128.52      straight up
           7
           58
           56
            25
            21
            42
            42
             0
```

```
                12.35        Move decimal point 2
22.    5.26^)64.96^10         places to the right
              52 6            in both divisor and
              12 36           dividend
              10 52
               1 84 1
               1 57 8
                 26 30        Zero is added to
                 26 30        dividend and brought
                     0        down
```

```
                110.5         Move decimal point 3
23.    .025^)2.762^5           places to the right
              2 5             in both divisor and
               26             dividend
               25
                12 5
                12 5
                    0
```

```
                 .4
24.   2/5    5)2.0
              2 0
                0

      Answer:  .4
```

```
                 .875
25.   7/8    8)7.000
              6 4
                60
                56
                 40
                 40
                  0

      Answer:  .875
```

```
                 .85
26.   17/20  20)17.00
              16 0
                1 00
                1 00
                   0

      Answer:  .85
```

27. $\dfrac{12}{14}$ $\dfrac{12}{14} = \dfrac{6}{7}$ $7\overline{)6.0000}.8571$
$\underline{5\ 6}$
40
$\underline{35}$
50
$\underline{49}$
10
$\dfrac{7}{3}$ Stop

Answer: .857

28. $\dfrac{3\tfrac{1}{2}}{7} = \dfrac{\tfrac{7}{2}}{\tfrac{7}{1}} = \dfrac{7}{2} \div \dfrac{7}{1} = \dfrac{7}{2} \cdot \dfrac{1}{7} = \dfrac{1}{2}$

29. $\dfrac{6\tfrac{1}{4}}{12\tfrac{1}{2}} = \dfrac{\tfrac{25}{4}}{\tfrac{25}{2}} = \dfrac{25}{4} \div \dfrac{25}{2} = \dfrac{25}{4} \cdot \dfrac{2}{25} = \dfrac{1}{2}$

30. $\dfrac{1\tfrac{5}{8}}{13} = \dfrac{\tfrac{13}{8}}{\tfrac{13}{1}} = \dfrac{13}{8} \div \dfrac{13}{1} = \dfrac{13}{8} \cdot \dfrac{1}{13} = \dfrac{1}{8}$

31. $\dfrac{8 \text{ minutes}}{1 \text{ hour}} = \dfrac{8 \text{ minutes}}{1 \cdot 60 \text{ minutes}}$
$ = \dfrac{8}{60} = \dfrac{2}{15}$

32. $\dfrac{3 \text{ quarts}}{3 \text{ gallons}} = \dfrac{3 \text{ quarts}}{3 \cdot 4 \text{ quarts}}$
$ = \dfrac{3}{12} = \dfrac{1}{4}$

33. $\dfrac{5}{10} = \dfrac{10}{20}$
$\dfrac{5 \cdot 1}{5 \cdot 2} = \dfrac{1 \cdot 10}{2 \cdot 10}$
$\dfrac{1}{2} = \dfrac{1}{2}$ True

34. $\dfrac{16}{24} = \dfrac{2}{6}$
$\dfrac{8 \cdot 2}{8 \cdot 3} = \dfrac{1 \cdot 2}{3 \cdot 2}$
$\dfrac{2}{3}$ does not equal $\dfrac{1}{3}$.
False

35. $\dfrac{12}{14} = \dfrac{17}{21}$
$\dfrac{6 \cdot 2}{7 \cdot 2} = \dfrac{1 \cdot 17}{3 \cdot 7}$
$\dfrac{6}{7}$ does not equal $\dfrac{17}{21}$.
False

36. $\dfrac{63}{21} = \dfrac{48}{16}$
$\dfrac{21 \cdot 3}{21 \cdot 1} = \dfrac{3 \cdot 16}{1 \cdot 16}$
$\dfrac{3}{1} = \dfrac{3}{1}$ True

37. $\dfrac{8}{20} = \dfrac{40}{100}$
$8 \cdot 100 = 800$ Cross multiply
$40 \cdot 20 = 800$
True

38. $\dfrac{64}{144} = \dfrac{48}{108}$
$64 \cdot 108 = 6912$ Cross multiply
$144 \cdot 48 = 6912$
True

39. $\dfrac{1}{5} = \dfrac{t}{15}$
$5 \cdot t = 15$ Cross multiply
$\dfrac{5 \cdot t}{5} = \dfrac{15}{5}$
$t = 3$

40. $\dfrac{315}{45} = \dfrac{21}{z}$

$\dfrac{7}{1} = \dfrac{21}{z}$ Reduce

$\dfrac{7 \cdot z}{7} = \dfrac{21}{7}$ Cross multiply and divide by 7

$z = 3$

41. $\dfrac{9}{x} = \dfrac{57}{114}$

$\dfrac{9}{x} = \dfrac{1}{2}$ Reduce

$1 \cdot x = 9 \cdot 2$ Cross multiply

$x = 18$

42. $\dfrac{y}{30} = \dfrac{7\tfrac{1}{2}}{24}$

$24 \cdot y = 30 \cdot 7\tfrac{1}{2}$ Cross multiply

$24 \cdot y = \dfrac{30}{1} \cdot \dfrac{15}{2}$

$24 \cdot y = 225$

$\dfrac{24 \cdot y}{24} = \dfrac{225}{24}$

$y = 9\tfrac{3}{8}$

43. $\dfrac{7 \text{ watches}}{3 \text{ hours}} = \dfrac{x}{12 \text{ hours}}$

$3 \cdot x = 7 \cdot 12$

$3 \cdot x = 84$

$\dfrac{3 \cdot x}{3} = \dfrac{84}{3}$

$x = 28$ watches

44. $\dfrac{3\tfrac{1}{2} \text{ ounces}}{6 \text{ gallons}} = \dfrac{x}{102 \text{ gallons}}$

$6 \cdot x = 3\tfrac{1}{2} \cdot 102$

$6 \cdot x = \dfrac{7}{2} \cdot \dfrac{102}{1}$

$6 \cdot x = 357$

$\dfrac{6 \cdot x}{6} = \dfrac{357}{6}$

$x = 59\tfrac{1}{2}$ ounces

45. 25% = .25

46. 139.7% = 1.397

47. .025% = .00025

48. 2.62% = .0262

49. .68 = 68%

50. 2.71 = 271%

51. .023 = 2.3%

52. $12.5\% = \dfrac{12.5}{100} \cdot \dfrac{10}{10} = \dfrac{125}{1000} = \dfrac{1}{8}$

53. $37\tfrac{1}{2}\% = \dfrac{37\tfrac{1}{2}}{100} = \dfrac{75}{2} \div \dfrac{100}{1} = \dfrac{75}{2} \cdot \dfrac{1}{100} = \dfrac{3}{8}$

54. $\dfrac{7}{8} = \dfrac{P}{100}$

$8 \cdot P = 700$

$\dfrac{8 \cdot P}{8} = \dfrac{700}{8}$

$P = 87.5$

Answer: 87.5%

Cumulative Review 4-6 127

55. $\dfrac{1}{200} = \dfrac{P}{100}$

$200 \cdot P = 100$

$\dfrac{200 \cdot P}{200} = \dfrac{100}{200}$

$P = \dfrac{1}{2}$ or .5

Answer: $\dfrac{1}{2}$% or .5%

56. $\dfrac{50}{b} = \dfrac{5}{100}$

$\dfrac{50}{b} = \dfrac{1}{20}$

$1 \cdot b = 50 \cdot 20$

$b = 1000$

57. $\dfrac{a}{240} = \dfrac{10}{100}$

$\dfrac{a}{240} = \dfrac{1}{10}$

$10 \cdot a = 240$

$\dfrac{10 \cdot a}{10} = \dfrac{240}{10}$

$a = 24$

58. $\underset{P}{25\%}$ of $\underset{b}{240}$ is $\underset{a}{60}$.

$P = 25$

$b = 240$

$a = 60$

59. $\underset{a}{18}$ is $\underset{P}{\text{what percent}}$ of $\underset{b}{300}$?

$P =$ unknown

$b = 300$

$a = 18$

60. What $\underset{P}{\text{percent}}$ of $\underset{b}{272}$ (total people)

is $\underset{a}{204}$ (people who passed)?

$P =$ unknown

$b = 272$

$a = 204$

61. 25% of 500

$= .25 \cdot 500$

$= 125$

62. 5.4% of 1200

$= .054 \cdot 1200$

$= 64.8$

63. 150% of 614

$= 1.5 \cdot 614$

$= 921$

64. $\dfrac{78}{b} = \dfrac{40}{100}$

$\dfrac{78}{b} = \dfrac{2}{5}$

$2 \cdot b = 78 \cdot 5$

$\dfrac{2 \cdot b}{2} = \dfrac{390}{2}$

$b = 195$

65. $\dfrac{88}{b} = \dfrac{5\frac{1}{2}}{100}$

$5.5 \cdot b = 88 \cdot 100$

$\dfrac{5.5 \cdot b}{5.5} = \dfrac{8800}{5.5}$

$b = 1600$

128   Chapter 6   Percent

66. $\dfrac{55}{110} = \dfrac{P}{100}$

$\dfrac{1}{2} = \dfrac{P}{100}$

$2 \cdot P = 100$

$\dfrac{2 \cdot P}{2} = \dfrac{100}{2}$

$P = 50$

Answer: 50%

67. $\dfrac{36}{90} = \dfrac{P}{100}$

$\dfrac{2}{5} = \dfrac{P}{100}$

$5 \cdot P = 2 \cdot 100$

$\dfrac{5 \cdot P}{5} = \dfrac{200}{5}$

$P = 40$

Answer: 40%

68. Percent × base = amount

$P \times 28 = 7$

$\dfrac{P \cdot 28}{28} = \dfrac{7}{28}$

$P = .25$

Answer: 25%

69. Tax = $25 · 2%
    = 25 · .02
    = $.50

Total cost = 25 + .50 = $25.50

70. Tax = $196 · 7%
    = 196 · .07
    = $13.72

Total cost = 196 + 13.72 = $209.72

71. Commission = $14,622 · 5%
    = 14622 · .05
    = $731.10

72. Commission = $179,280 · 1.2%
    = 179,280 · .012
    = $2151.36

73. Discount = $76 · 35%
    = 76 · .35
    = $26.60

Amount paid = 76 − 26.60 = $49.40

74. Discount = $238.50 · 22.5%
    = 238.50 · .225
    = $53.66

Amount paid = 238.50 − 53.66
    = $184.84

75. 960 − 624 = 336, so 336 entrants is the amount of decrease; compare this to the original 960 entrants.

$\dfrac{336}{960} = \dfrac{P}{100}$

$960 \cdot P = 336 \cdot 100$

$\dfrac{960 \cdot P}{960} = \dfrac{33{,}600}{960}$

$P = 35$

Answer: 35%

76. 891 − 660 = 231, so 231 defects is the amount of increase; compare this to the original 660 defects.

$$\frac{231}{660} = \frac{P}{100}$$

$$660 \cdot P = 231 \cdot 100$$

$$\frac{660 \cdot P}{660} = \frac{23{,}100}{660}$$

$$P = 35$$

Answer: 35%

77. Interest = $357 · 18% · 2
    = 357 · .18 · 2
    = $128.52

    Total amount due = 357 + 128.52
                     = $485.52

78. Interest = $18,350 · 11% · $\frac{9}{12}$

    = 18,350 · .11 · $\frac{3}{4}$

    = $1513.88 (rounded)

    Total amount due = 18,350 + 1513.88
                     = $19,863.88

CHAPTER 7 MEASUREMENT

Section 7.1

3. 2 pints = 1 quart

7. 2000 pounds = 1 ton

11. 8 quarts · $\dfrac{1 \text{ gallon}}{4 \text{ quarts}} = \dfrac{8}{4} = 2$ gallons

15. 18 inches · $\dfrac{1 \text{ foot}}{12 \text{ inches}} = \dfrac{18}{12}$
$= 1.5$ or $1\dfrac{1}{2}$ feet

19. 136 ounces · $\dfrac{1 \text{ pound}}{16 \text{ ounces}}$
$= \dfrac{136}{16}$
$= 8.5$ or $8\dfrac{1}{2}$ pounds

23. $5\dfrac{1}{2}$ pounds · $\dfrac{16 \text{ ounces}}{1 \text{ pound}} = 5\dfrac{1}{2} \cdot 16$
$= 88$ ounces

27. 2 miles · $\dfrac{5280 \text{ feet}}{1 \text{ mile}} = 2 \cdot 5280$
$= 10{,}560$ feet

31. 17 pints · $\dfrac{1 \text{ quart}}{2 \text{ pints}} = 17 \cdot \dfrac{1}{2}$
$= 8.5$ or $8\dfrac{1}{2}$ quarts

17 pints · $\dfrac{1 \text{ quart}}{2 \text{ pints}} \cdot \dfrac{1 \text{ gallon}}{4 \text{ quarts}}$
$= 17 \cdot \dfrac{1}{2} \cdot \dfrac{1}{4} = \dfrac{17}{8}$
$= 2.125$ or $2\dfrac{1}{8}$ gallons

35. 21,120 feet · $\dfrac{1 \text{ mile}}{5280 \text{ feet}}$
$= \dfrac{21{,}120}{5280} = 4$ miles

39. $7\dfrac{1}{8}$ tons · $\dfrac{2000 \text{ pounds}}{1 \text{ ton}} = 7\dfrac{1}{8} \cdot 2000$
$= 14{,}250$ pounds

43. $8\dfrac{1}{4}$ hours · $\dfrac{60 \text{ minutes}}{1 \text{ hour}} \cdot \dfrac{60 \text{ seconds}}{1 \text{ minute}}$
$= 8\dfrac{1}{4} \cdot 60 \cdot 60$
$= \dfrac{33}{\cancel{4}} \cdot \dfrac{\overset{15}{\cancel{60}}}{1} \cdot \dfrac{60}{1}$
$= 29{,}700$ seconds

47. 2 weeks ___ 15 days

2 weeks · $\dfrac{7 \text{ days}}{1 \text{ week}}$

14 days $<$ 15 days

2 weeks $\underline{<}$ 15 days

51. 32 days ___ 4 weeks

4 weeks · $\dfrac{7 \text{ days}}{1 \text{ week}}$

32 days $>$ 28 days

32 days $\underline{>}$ 4 weeks

Section 7.2

3. 1 pound 20 ounces

    20 ounces = $\frac{20 \text{ ounces}}{16 \text{ ounces in a pound}}$

    = 1 pound 4 ounces

    1 pound + 1 pound 4 ounces

    = 2 pounds 4 ounces

7. 1 mile 7250 feet

    7250 feet = $\frac{7250 \text{ feet}}{5280 \text{ feet in a mile}}$

    = 1 mile 1970 feet

    1 mile + 1 mile 1970 feet

    = 2 miles 1970 feet

11. 7 gallons 6 quarts

    6 quarts = $\frac{6 \text{ quarts}}{4 \text{ quarts in a gallon}}$

    = 1 gallon 2 quarts

    7 gallons + 1 gallon 2 quarts

    = 8 gallons 2 quarts

15. 2 weeks 9 days <u>60 hours</u>

    = 2 weeks 9 days + 2 days 12 hours

    = 2 weeks <u>11 days</u> 12 hours

    = 2 weeks 1 week 4 days + 12 hours

    = 3 weeks 4 days 12 hours

19.   8 feet 3 inches
    + 2 feet 10 inches
    ──────────────────
     10 feet <u>13 inches</u>

    = 10 feet + 1 foot 1 inch

    = 11 feet 1 inch

23.   7 pounds 12 ounces
    + 3 pounds 10 ounces
    ────────────────────
     10 pounds <u>22 ounces</u>

    = 10 pounds + 1 pound 6 ounces

    = 11 pounds 6 ounces

27.   2 gallons 3 quarts
    − 1 gallon 2 quarts
    ────────────────────
      1 gallon 1 quart

31. 2 3+1=4
      $\cancel{3}$ yards $\cancel{1}$ foot
    − 2 yards 2 feet
    ─────────────────
              2 feet

35.   6 feet 4 inches
    ×   2
    ────────────────
     12 feet 8 inches

39.   75 yards 2 feet
    ×   10
    ─────────────────
     750 yards <u>20 feet</u>

    = 750 yards + 6 yards 2 feet

    = 756 yards 2 feet

43. 2 gallons 1 quart
    2)4 gallons 2 quarts
     4 2
     ─ ─
     0 0

47. 3 yards 1 foot
    2)6 yards 2 feet
     6 2
     ─ ─
     0 0

51.
```
  3 gallons 3 quarts 1 pint  1 cup
+ 2 gallons 2 quarts 1 pint  1 cup
  5 gallons 5 quarts 2 pints 2 cups
```
= 5 gallons 5 quarts 2 pints + 1 pint

= 5 gallons 5 quarts 3 pints

= 5 gallons 5 quarts + 1 quart 1 pint

= 5 gallons 6 quarts 1 pint

= 5 gallons + 1 gallon 2 quarts 1 pint

= 6 gallons 2 quarts 1 pint

55. .6071

Hundredths: 0 (second digit to the right of the decimal point)

Thousandths: 7 (third digit to the right of the decimal point)

Section 7.3

3. 1 m = 1000 mm

7. answer varies — about 8 cm

11. $8 \text{ m} \cdot \frac{1000 \text{ mm}}{1 \text{ m}} = 8 \cdot 1000 = 8000 \text{ mm}$

15. $52.5 \text{ cm} \cdot \frac{1 \text{ m}}{100 \text{ cm}} = \frac{52.5}{100} = .525 \text{ m}$

19. $118.6 \text{ cm} \cdot \frac{10 \text{ mm}}{1 \text{ cm}} = 118.6 \cdot 10$
$= 1186 \text{ mm}$

23. $27{,}500 \text{ m} \cdot \frac{1 \text{ km}}{1000 \text{ m}} = \frac{27{,}500}{1000}$
$= 27.5 \text{ km}$

27. $82 \text{ cm} \cdot \frac{10 \text{ mm}}{1 \text{ cm}} = 82 \cdot 10 = 820 \text{ mm}$

31. $82 \text{ cm} \cdot \frac{1 \text{ m}}{100 \text{ cm}} = \frac{82}{100} = .82 \text{ m}$ or smaller than one meter.

35. $37{,}031.6 \text{ mm} \cdot \frac{1 \text{ m}}{1000 \text{ mm}} \cdot \frac{1 \text{ km}}{1000 \text{ m}}$
$= \frac{37{,}031.6}{1{,}000{,}000}$
$= .0370316 \text{ km}$

39. $.08 = \frac{8}{100} = \frac{2 \cdot 4}{25 \cdot 4} = \frac{2}{25}$

Section 7.4

3. $8.7 \ell \cdot \frac{1000 \text{ ml}}{1 \ell} = 8.7 \cdot 1000$
$= 8700 \text{ ml}$

7. $8974 \text{ ml} \cdot \frac{1 \ell}{1000 \text{ ml}} = \frac{8974}{1000} = 8.974 \ \ell$

11. $8.64 \text{ kl} \cdot \frac{1000 \ \ell}{1 \text{ kl}} = 8.64 \cdot 1000$
$= 8640 \ \ell$

15. $5.2 \text{ kg} \cdot \frac{1000 \text{ g}}{1 \text{ kg}} = 5.2 \cdot 1000 = 5200 \text{ g}$

19. $7634 \text{ cg} \cdot \frac{1 \text{ g}}{100 \text{ cg}} = \frac{7634}{100} = 76.34 \text{ g}$

23. This is unreasonable since 4 liters of water is about 1 gallon, while 3 milligrams of Epsom salts is only about as much as 3 grains of sand. That is very few Epsom salts.

7.5 Metric to English Conversions   133

27. This is unreasonable since

$94.3 \text{ ml} \cdot \frac{1 \ell}{1000 \text{ ml}} = \frac{94.3}{1000} = .0943 \ \ell,$

which is about 1/10 of a liter, which is about 1/10 of a quart or 1/10 of 32 ounces, or about 3 ounces.

31. $1 \ \ell \cdot \frac{1000 \text{ ml}}{1 \ \ell} = 1000 \text{ ml}$

Each milliliter of helium weighs .0002 grams, so to find the weight of $1 \ \ell = 1000$ ml, multiply .0002 by 1000.

$.0002 \cdot 1000 = .2 \text{ grams}$

35. 1.6093   4 decimal places
    ×   24   + 0 decimal places
    ─────
     6 4372   4 decimal places
    32 186   in product
    ─────
    38.6232

Section 7.5

3. $80 \text{ m} \cdot 3.28 = 262.4$ feet

7. $982 \text{ yards} \cdot .9144 = 897.9$ m (rounded)

11. $28.6 \ \ell \cdot 1.06 = 30.3$ quarts (rounded)

19. Perimeter = $2l + 2w$
    $= 2 \cdot 22 + 2 \cdot 31$
    $= 106$ cm or 1.06 m of wood needed

    1.06 meters $\cdot 3.28 = 3.4768$ feet
    3.4768 feet $\cdot \$1.20$ per foot = \$4.17

23. $C = \frac{5(68 - 32)}{9}$

    $= \frac{5(36)}{9}$

    $= 20°C$

27. $C = \frac{5(98 - 32)}{9}$

    $= \frac{5(66)}{9}$

    $= 37°C$

31. $F = \frac{9 \cdot 25}{5} + 32$

    $= 45 + 32$

    $= 77°F$

35. $C = \frac{5(136 - 32)}{9}$

    $= \frac{5(104)}{9}$

    $= 58°C$

39. $2 \cdot 8 + 2 \cdot 8$
    $= 16 + 2 \cdot 8$ *Perform leftmost multiplication*
    $= 16 + 16$ *Perform other multiplication*
    $= 32$ *Add last*

43. $(5^2) + (4^2)$
    $= 25 + 16$ *Simplify exponents*
    $= 41$ *Add last*

Chapter 7 Review Exercises

1. 1 pound = 16 ounces

2. 3 feet = 1 yard

3. 1 ton = 2000 pounds

4. 4 quarts = 1 gallon

5. 5280 feet = 1 mile

6. 1 yard · $\frac{3 \text{ feet}}{1 \text{ yard}}$ · $\frac{12 \text{ inches}}{1 \text{ foot}}$
    = 1 · 3 · 12 = 36 inches

7. 12 feet · $\frac{1 \text{ yard}}{3 \text{ feet}}$ = $\frac{12}{3}$ = 4 yards

8. 7 days · $\frac{24 \text{ hours}}{1 \text{ day}}$ = 7 · 24 = 168 hours

9. 4 miles · $\frac{5280 \text{ feet}}{1 \text{ mile}}$ = 4 · 5280
    = 21,120 feet

10. 1 foot 19 inches
    19 inches = $\frac{19 \text{ inches}}{12 \text{ inches in one foot}}$
    = 1 foot 7 inches
    1 foot + 1 foot 7 inches
    = 2 feet 7 inches

11. 2 miles 6000 feet
    6000 feet = $\frac{6000 \text{ feet}}{5280 \text{ feet in one mile}}$
    = 1 mile 720 feet
    2 miles + 1 mile 720 feet
    = 3 miles 720 feet

12. 2 yards 7 feet 15 inches
    = 2 yards 7 feet + 1 foot 3 inches
    = 2 yards 8 feet 3 inches
    = 2 yards + 2 yards 2 feet + 3 inches
    = 4 yards 2 feet 3 inches

13. 2 quarts 3 pints 1 cup
    = 2 quarts + 1 quart 1 pint + 1 cup
    = 3 quarts 1 pint 1 cup

14.   5 feet 7 inches
    + 2 feet 8 inches
      7 feet 15 inches
    = 7 feet + 1 foot 3 inches
    = 8 feet 3 inches

15.   7   16+6=22
        8̸ pounds 6̸ ounces
    − 3 pounds 8 ounces
        4 pounds 14 ounces

16.   5 pounds 10 ounces
    ×   5
    25 pounds 50 ounces
    = 25 pounds + 3 pounds 2 ounces
    = 28 pounds 2 ounces

17.   5 feet 4 inches
    2)10 feet 8 inches
        10   8
         0   0

18. 1 m = 100 cm

19. 1 km = 1000 m

20. 8 m · $\frac{100 \text{ cm}}{1 \text{ m}}$ = 8 · 100 = 800 cm

21. $3781 \text{ mm} \cdot \dfrac{1 \text{ m}}{1000 \text{ mm}} = \dfrac{3781}{1000} = 3.781 \text{ m}$

22. $.056 \text{ m} \cdot \dfrac{1000 \text{ mm}}{1 \text{ m}} = .056 \cdot 1000$
    $= 56 \text{ mm}$

23. $1.27 \text{ km} \cdot \dfrac{1000 \text{ m}}{1 \text{ km}} = 1.27 \cdot 1000$
    $= 1270 \text{ m}$

24. $3 \ell \cdot \dfrac{100 \text{ cl}}{1 \ell} = 3 \cdot 100 = 300 \text{ cl}$

25. $680 \text{ g} \cdot \dfrac{1 \text{ kg}}{1000 \text{ g}} = \dfrac{680}{1000} = .68 \text{ kg}$

26. $485 \text{ cg} \cdot \dfrac{10 \text{ mg}}{1 \text{ cg}} = 485 \cdot 10 = 4850 \text{ mg}$

27. $17.6 \text{ kl} \cdot \dfrac{1000 \ell}{1 \text{ kl}} = 17.6 \cdot 1000$
    $= 17{,}600 \ \ell$

28. $5.3 \text{ g} \cdot \dfrac{100 \text{ cg}}{1 \text{ g}} = 5.3 \cdot 100 = 530 \text{ cg}$

29. $5000 \text{ g} \cdot \dfrac{1 \text{ kg}}{1000 \text{ g}} = \dfrac{5000}{1000} = 5 \text{ kg}$

30. $10 \text{ m} \cdot 1.094 = 10.9 \text{ yards}$

31. $1.4 \text{ m} \cdot 39.37 = 55.1 \text{ inches}$

32. $108 \text{ km} \cdot .624 = 67.4 \text{ miles}$

33. $800 \text{ miles} \cdot 1.6093 = 1287.4 \text{ km}$

34. $23 \text{ quarts} \cdot .946 = 21.8 \ \ell$

35. $41.5 \ \ell \cdot 1.06 = 44.0 \text{ quarts}$

36. $55 \cdot 1.6093 = 88.5$ kilometers per hour (rounded)

37. $\$6.95 \cdot 1.094 = \$7.60$ per meter

38. $C = \dfrac{5(77 - 32)}{9}$
    $= \dfrac{5(45)}{9}$
    $= 25°C$

39. $C = \dfrac{5(176 - 32)}{9}$
    $= \dfrac{5(144)}{9}$
    $= 80°C$

40. $C = \dfrac{5(92 - 32)}{9}$
    $= \dfrac{5(60)}{9}$
    $= 33.3°C$

41. $C = \dfrac{5(159 - 32)}{9}$
    $= \dfrac{5(127)}{9}$
    $= 70.6°C$

42. $F = \dfrac{9 \cdot 50}{5} + 32$
    $= 90 + 32$
    $= 122°F$

43. $F = \dfrac{9 \cdot 280}{5} + 32$
    $= 504 + 32$
    $= 536°F$

136   Chapter 7   Measurement

44. $F = \dfrac{9 \cdot 100}{5} + 32$

  $= 180 + 32$

  $= 212°F$

45. $C = \dfrac{5(180 - 32)}{9}$

  $= \dfrac{5(148)}{9}$

  $= 82.2°C$

46.
```
   5 quarts  2 pints
 + 1 quart   1 pint
   6 quarts  3 pints
```
  $= 6$ quarts $+ 1$ quart 1 pint

  $= 7$ quarts 1 pint

47.
```
       7+4=11
    6    ⁊  24+3=27
    7̸ weeks 5̸ days 3̸ hours
  - 2 weeks 6 days  9 hours
    4 weeks 5 days 18 hours
```

48.
```
      5 weeks 2 days  9 hours
   ×                        4
     20 weeks 8 days 36 hours
```
  $= 20$ weeks 8 days $+ 1$ day 12 hours

  $= 20$ weeks 9 days 12 hours

  $= 20$ weeks $+ 1$ week 2 days $+ 12$ hours

  $= 21$ weeks 2 days 12 hours

49.
```
        1 gallon   1 quart    1 pint
   5) 6 gallons   2 quarts   3 pints
      5          + 4 quarts + 2 pints
      1 gallon    6 quarts    5 pints
                  5           5
                  1 quart     0
```

50. 2.75 cm $\cdot \dfrac{10 \text{ mm}}{1 \text{ cm}} = 2.75 \cdot 10 = 27.5$ mm

51. $42{,}885$ mm $\cdot \dfrac{1 \text{ m}}{1000 \text{ mm}} = \dfrac{42{,}885}{1000}$

  $= 42.885$ m

52. 1.3 km $\cdot \dfrac{100{,}000 \text{ cm}}{1 \text{ km}} = 1.3 \cdot 100{,}000$

  $= 130{,}000$ cm

53. 7835 mg $\cdot \dfrac{1 \text{ cg}}{10 \text{ mg}} = \dfrac{7835}{10} = 783.5$ cg

54. $67{,}500$ mg $\cdot \dfrac{1 \text{ kg}}{1{,}000{,}000 \text{ mg}}$

  $= \dfrac{67{,}500}{1{,}000{,}000} = .0675$ kg

55. $.0894$ kl $\cdot \dfrac{1000 \ \ell}{1 \text{ kl}} = .0894 \cdot 1000$

  $= 89.4 \ \ell$

56. 5 pounds $\cdot \dfrac{16 \text{ ounces}}{1 \text{ pound}} = 5 \cdot 16$

  $= 80$ ounces

57. 90 inches $\cdot \dfrac{1 \text{ foot}}{12 \text{ inches}} = \dfrac{90}{12} = 7\tfrac{1}{2}$ feet

58. 364 days $\cdot \dfrac{1 \text{ week}}{7 \text{ days}} = \dfrac{364}{7} = 52$ weeks

59. 2 days 25 hours 150 minutes

  $= 2$ days 25 hours $+ 2$ hours 30 minutes.

  $= 2$ days 27 hours 30 minutes.

  $= 2$ days $+ 1$ day 3 hours $+ 30$ minutes.

  $= 3$ days 3 hours 30 minutes.

60. 2 miles 11,350 feet 110 inches

    = 2 miles 11,350 feet + 9 feet 2 inches

    = 2 miles 11,359 feet 2 inches

    = 2 miles + 2 miles 799 feet + 2 inches

    = 4 miles 799 feet 2 inches

61. 24 feet • .305 = 7.3 m

62. 5.8 • .9144 = 5.3 m

63. 52.3 gallons • 3.785 = 198.0 ℓ

64. 3.5 pounds • 454 = 1589 grams

65. $2.00 • .264 = $.528 or 52.8¢

66. 1000 liters • .264 = 264 gallons

67. 100 yards • .9144 = 91.4 meters

68. $C = \frac{5(5000 - 32)}{9}$

    $= \frac{5(4968)}{9}$

    $= 2760°C$

Chapter 7 Test

1. 1 gallon = 4 quarts

2. 36 feet $\cdot \frac{1 \text{ yard}}{3 \text{ feet}} = \frac{36}{3} = 12$ yards

3. 228 hours $\cdot \frac{1 \text{ day}}{24 \text{ hours}} = \frac{228}{24} = 9.5$ days

4. 42 pints $\cdot \frac{1 \text{ gallon}}{8 \text{ pints}} = \frac{42}{8} = 5\frac{1}{4}$ gallons

5. 128 ounces $\cdot \frac{1 \text{ pound}}{16 \text{ ounces}} = \frac{128}{16}$

    = 8 pounds

6. 38,016 feet $\cdot \frac{1 \text{ mile}}{5280 \text{ feet}} = \frac{38,016}{5280}$

    $= 7.2$ or $7\frac{1}{5}$ miles

7. 9 feet 4 inches
   +7 feet 9 inches
   ‾‾‾‾‾‾‾‾‾‾‾‾‾‾‾‾
   16 feet 13 inches

    = 16 feet + 1 foot 1 inch

    = 17 feet 1 inch

8. 5 gallons $\overset{2}{\cancel{3}}$ quarts $\overset{3}{\cancel{1}}$ pint
   − 2 gallons 2 quarts 2 pints
   ‾‾‾‾‾‾‾‾‾‾‾‾‾‾‾‾‾‾‾‾‾‾‾‾‾‾‾‾
     3 gallons   1 pint

9.   5 pounds 13 ounces
    ×   6
    ‾‾‾‾‾‾‾‾‾‾‾‾‾‾‾‾‾‾‾‾
     30 pounds 78 ounces

    = 30 pounds + 4 pounds 14 ounces

    = 34 pounds 14 ounces

10.   1 yard 2 feet 2 inches
    4)6 yards 2 feet 8 inches
      4   +6 feet 8
      ‾   ‾‾‾‾‾‾‾ ‾
      2 yards 8 feet 0
                      8
                      ‾
                      0

11. 250 cm $\cdot \frac{1 \text{ m}}{100 \text{ cm}} = \frac{250}{100} = 2.5$ m

138   Chapter 7   Measurement

12. $4.6 \cdot \frac{100 \text{ cm}}{1 \text{ m}} = 4.6 \cdot 100 = 460$ cm

13. $.4 \text{ km} \cdot \frac{100,000 \text{ cm}}{1 \text{ km}} = .4 \cdot 100,000$
    $= 40,000$ cm

14. $8412 \text{ g} \cdot \frac{1 \text{ kg}}{1000 \text{ g}} = \frac{8412}{1000} = 8.412$ kg

15. $9 \ell \cdot \frac{100 \text{ cl}}{1 \ell} = 9 \cdot 100 = 900$ cl

16. $15.6 \text{ cl} \cdot \frac{10 \text{ ml}}{1 \text{ cl}} = 15.6 \cdot 10 = 156$ ml

17. $198 \text{ m} \cdot \frac{1 \text{ km}}{1000 \text{ m}} = \frac{198}{1000}$
    $= .198$ km

18. $2.61 \text{ kg} \cdot \frac{1000 \text{ g}}{1 \text{ kg}} = 2.61 \cdot 1000$
    $= 2610$ g

19. $18.4 \text{ feet} \cdot .305 = 5.6$ meters

20. $72.1 \text{ pounds} \cdot .454 = 32.7$ kilograms

21. $948 \text{ grams} \cdot .0022 = 2.1$ pounds

22. $6.4 \text{ miles} \cdot 1.6093 = 10.3$ kilometers

23. $.33 per liter
    $= .33 \frac{\text{dollars}}{\text{liter}} \cdot \frac{1 \text{ liter}}{.264 \text{ gallons}}$
    $= \frac{.33}{.264} \frac{\text{dollars}}{\text{gallon}}$
    $= 1.25 \frac{\text{dollars}}{\text{gallon}}$
    $= \$1.25$ per gallon

24. $7.7 \text{ pounds} \cdot .454 = 3.5$ kg

25. $C = \frac{5(76 - 32)}{9}$
    $= \frac{5(44)}{9}$
    $= 24.4°C$

26. $F = \frac{9 \cdot 172}{5} + 32$
    $= \frac{1548}{5} + 32$
    $= 309.6 + 32$
    $= 341.6°F$

CHAPTER 8 GEOMETRY
Section 8.1

3. The figure is a ray since it is a part of a line that has one endpoint and continues forever in one direction.

7. The lines are perpendicular since they intersect at right angles.

11. The lines are parallel since they do not intersect.

15. The name of the angle is ∠POS or ∠SOP. Note that the vertex O appears as the middle letter in the name of the angle; it would not be correct to call the angle simply ∠O, however, because more than just this one angle have point O as their vertex.

19. The angle is a right angle, as indicated by the box drawn between the two sides.

23. The angle is a straight angle since it is a straight line; it is made up of two right angles, so its measure is 2(90°) = 180°.

27. False
∠SQP is labelled as a right angle, so its measure is equal to 90°, not greater than 90°.

31. $\overset{1}{3}6$
    $\underline{+\ 45}$
    81

35. 63 − 37 + 43
    = 26 + 43   *Subtract first*
    = 69   *Add last*

Section 8.2

3. The complement of 65° is 25° since
$$90° - 65° = 25°.$$

7. The supplement of 30° is 150° since
$$180° - 30° = 150°.$$

11. The pairs of supplementary angles are as follows:

    ∠COA and ∠AOD since
    $$155° + 25° = 180°;$$
    ∠COA and ∠COF since
    $$155° + 25° = 180°;$$
    ∠COF and ∠DOF since
    $$25° + 155° = 180°;$$
    ∠DOF and ∠AOD since
    $$155° + 25° = 180°.$$

15. Since ∠COE and ∠GOH are vertical angles, they are congruent. This means they are equivalent in size.
$$∠GOH = 63°$$

19. Since ∠BOD and ∠AOC are vertical angles, they are congruent.
$$∠AOC = 54°$$

23. Since ∠ABC and ∠BCD are equivalent in size, they are congruent.

$$\angle ABC \cong \angle BCD$$

Since ∠ABC and ∠ABE are supplementary, ∠ABE is 47°.

$$180° - 133° = 47°$$

Since ∠DCG and ∠BCD are supplementary, ∠DCG is 47°.

Therefore, since ∠ABE and ∠DCG are equivalent in size (47°),

$$\angle ABE \cong \angle DCG.$$

27. 81 ÷ 9 · 4   *Divide first*
        9 · 4   *Multiply*
          36

Section 8.3

3. P = 2l + 2w
    = 2 · 14 + 2 · 5
    = 28 + 10
    = 38 ft

    A = lw
    = 14 · 5
    = 70 ft²

7. P = 2l + 2w
    = 2 · 21.2 + 2 · 12.9
    = 42.4 + 25.8
    = 68.2 m

    A = lw
    = 21.2 · 12.9
    = 273.48 m²

11. P = 4s
    = 4 · 8.4
    = 33.6 yd

    A = s²
    = 8.4²
    = 70.56 yd²

15. See drawing in textbook.

    P = 28 + 17 + 12 + 4 + 4 + 4 + 12 + 17
    = 98 m
    A (left) = 17 · 28 = 476 m²
    A (right) = 4² = 16 m²
    A (total) = 476 + 16 = 492 m²

19. A = l · w
    = 30 · 24
    = 720 in²

23. Cost = 1472 · .96
    = $1413.12

27. $30 \text{ cm} \cdot \dfrac{1 \text{ m}}{100 \text{ cm}} = \dfrac{30}{100} = .3 \text{ m}$

31. $2700 \text{ inches} \cdot \dfrac{1 \text{ foot}}{12 \text{ inches}} \cdot \dfrac{1 \text{ yard}}{3 \text{ feet}}$

    $= \dfrac{2700}{36} = 75 \text{ yards}$

Section 8.4

3. $P = 51.8 + 51.8 + 51.8 + 51.8$
    or $4 \cdot 51.8$
    $= 207.2$ m

7. $A = bh \quad b = 41, h = 24$
    $= 41 \cdot 24$
    $= 984$ yd²

11. $A = \frac{1}{2}h(b + B) \quad h = 42, b = 61.4, B = 86.2$
    $= \frac{1}{2} \cdot 42 \ (61.4 + 86.2)$
    $= 21 \ (147.6)$
    $= 3099.6$ cm²

15. $A = \frac{1}{2}h(b + B) \quad h = 45, b = 80, B = 110$
    $A = \frac{1}{2} \cdot 45 \cdot (80 + 110)$
    $= \frac{1}{2} \cdot 45(190)$
    $= \frac{1}{\cancel{2}} \cdot \frac{45}{1} \cdot \frac{\cancel{190}\,^{95}}{1}$
    $= 45 \cdot 95 = 4275$ ft²
    Cost $= 4275 \cdot .22 = \$940.50$

19. A (top) $= 96 \cdot 96 = 9216$ ft²
    A (middle) $= 72 \cdot 96 = 6912$ ft²
    A (bottom) $= 96 \cdot 96 = 9216$ ft²
    A (total) $= 9216 + 6912 + 9216$
    $= 25,344$ ft²

23. $25 \cdot \frac{50}{7} \cdot \frac{28}{75}$
    $= \frac{25}{1} \cdot \frac{\cancel{50}\,^2}{\cancel{7}\,_1} \cdot \frac{\cancel{28}\,^4}{\cancel{75}\,_3}$
    $= \frac{25 \cdot 2 \cdot 4}{1 \cdot 1 \cdot 3}$
    $= \frac{200}{3}$

Section 8.5

3. $P = 11 + 17 + 9 = 37$ yd

7. A (triangle) $= \frac{1}{2}bh \quad h = 9, b = 12$
    $= \frac{1}{2} \cdot 9 \cdot 12$
    $= 54$ m²
    A (square) $= s^2 \quad s = 12$
    $= 12^2$
    $= 144$ m²
    A (total) $= 54 + 144 = 198$ m²

11. A (trapezoid) $= \frac{1}{2}h(b + B)$
    $= \frac{1}{2} \cdot 24(28 + 57)$
    $= 12(85)$
    $= 1020$ cm²
    A (triangle) $= \frac{1}{2}bh$
    $= \frac{1}{2} \cdot 19 \cdot 16$
    $= 152$ cm²
    A (shaded) $= 1020 - 152 = 868$ cm²

15. $\begin{array}{r} 3.14 \\ \times16 \\ \hline 18\;84 \\ 31\;4 \\ \hline 50.24 \end{array}$ 2 decimal places
+ 0 decimal places
2 decimal places in product

Section 8.6

3. $r = \dfrac{d}{2}$

   $= \dfrac{58.9}{2}$

   $= 29.45$ m

7. $C = \pi d$

   $= 3.14 \cdot 50$

   $= 157$ m

   $A = \pi r^2 \qquad r = \dfrac{50}{2} = 25$

   $= 3.14 \cdot 25^2$

   $= 3.14 \cdot 625$

   $= 1962.5$ m²

11. $C = \pi d$

    $= 3.14 \cdot 7\tfrac{1}{2}$ or $3.14 \cdot 7.5$

    $= 23.6$ m (rounded)

    $A = \pi r^2 \qquad r = \dfrac{7.5}{2} = 3.75$

    $= 3.14 \cdot 3.75^2$

    $= 3.14 \cdot 14.0625$

    $= 44.2$ m² (rounded)

15. A (semicircle) $= \tfrac{1}{2}\pi r^2$

    $= \tfrac{1}{2} \cdot 3.14 \cdot 10^2$

    $= \tfrac{1}{2} \cdot 3.14 \cdot 100$

    $= 157$ cm²

   A (triangle) $= \tfrac{1}{2}bh$

   $= \tfrac{1}{2} \cdot 20 \cdot 10$

   $= 100$ cm²

   A (shaded) $= 157 - 100 = 57$ cm²

19. $C = 2\pi r = 2 \cdot 3.14 \cdot 2 = 12.6$ m

23. A (circle) $= \pi r^2$

    $= 3.14 \cdot 8^2$

    $= 3.14 \cdot 64$

    $= 200.96$

    A (triangle) $= \tfrac{1}{2}bh$

    $= \tfrac{1}{2} \cdot 16 \cdot 8$

    $= 64$ ft² × 2 triangles

    $= 128$ ft²

    A (shaded) $= 200.96 - 128$

    $= 73.0$ ft² (rounded)

27. $\dfrac{6}{7} \div \dfrac{12}{21}$

    $= \dfrac{\cancel{6}^{1}}{\cancel{7}_{1}} \cdot \dfrac{\cancel{21}^{3}}{\cancel{12}_{2}}$

    $= \dfrac{1 \cdot 3}{1 \cdot 2}$

    $= \dfrac{3}{2}$

Section 8.7

3. $V = lwh$
    $= 12 \cdot 12 \cdot 12$
    $= 1728$ in³

7. $V = \frac{2}{3}\pi r^3$
    $= \frac{2}{3} \cdot 3.14 \cdot 12^3$
    $= \frac{2}{3} \cdot 3.14 \cdot 1728$
    $= \frac{2}{3} \cdot 5425.92$
    $= 10851.84 \div 3$ *Divide by 3 last*
    $= 3617.3$ in³ (rounded)

11. $A = \pi r^2 h$
    $= 3.14 \cdot 18^2 \cdot 5$
    $= 3.14 \cdot 324 \cdot 5$
    $= 5086.8$ ft³

15. $V = \frac{1}{3}bh$
    $b = 7 \cdot 5 = 35$ m²
    $V = \frac{1}{3} \cdot 35 \cdot 15 = 175$ m³

19. $r = 6.3$ cm, $h = 15.8$ cm;
    $V = \pi r^2 h$
    $= 3.14 \cdot 6.3 \cdot 6.3 \cdot 15.8$
    $= 1969.1003$ Round to 1969.1 cm³.

23. V (large cylinder) $= \pi r^2 h$
    $= 3.14 \cdot 4^2 \cdot 8$
    $= 401.92$ m³
    V (small cylinder) $= \pi r^2 h$
    $= 3.14 \cdot 2^2 \cdot 8$
    $= 100.48$ m³
    V (shaded) $= 401.92 - 100.48$
    $= 301.4$ m³

27. $\quad\overset{1}{}$
    $\quad 47.123$
    $+ 36.345$
    $\overline{\quad 83.468}$

Section 8.8

3. $\sqrt{64} = \sqrt{8 \cdot 8} = 8$

7. $\sqrt{45} = 6.708$

11. $\sqrt{106} = 10.296$

15. $\sqrt{173} = 13.153$

19. Hypotenuse $= \sqrt{5^2 + 12^2}$
    $= \sqrt{25 + 144}$
    $= \sqrt{169}$
    $= 13$ cm

23. leg $= \sqrt{10^2 - 6^2}$
    $= \sqrt{100 - 36}$
    $= \sqrt{64}$
    $= 8$ in

27. Hypotenuse $= \sqrt{4^2 + 7^2}$
    $= \sqrt{16 + 49}$
    $= \sqrt{65}$
    $= 8.062$ in

31. Hypotenuse $= \sqrt{1.8^2 + 3.4^2}$
    $= \sqrt{3.24 + 11.56}$
    $= \sqrt{14.8}$ or 3.847 in

35. leg $= \sqrt{24.1^2 - 14.9^2}$
    $= \sqrt{580.81 - 222.01}$
    $= \sqrt{358.8}$ or 18.942 in

39. Hypotenuse $= \sqrt{11^2 + 15^2}$
    $= \sqrt{121 + 225}$
    $= \sqrt{346}$ or 18.601 mi

43. $AC = \sqrt{9^2 + (7+5)^2}$
    $= \sqrt{81 + 12^2}$
    $= \sqrt{81 + 144}$
    $= \sqrt{225} = 15$
    $BC = AC - AB = 15 - 7 = 8$ ft
    $BD = \sqrt{8^2 - 5^2}$
    $= \sqrt{64 - 25}$
    $= \sqrt{39}$
    $= 6.245$ ft (rounded)
    Answer: $BC = 8$ ft, $BD = 6.245$ ft

47. $\dfrac{2\frac{1}{2}}{3\frac{1}{3}} = \dfrac{x}{1\frac{1}{4}}$

    $3\frac{1}{3} \cdot x = 2\frac{1}{2} \cdot 1\frac{1}{4}$

    $\dfrac{10}{3} \cdot x = \dfrac{5}{2} \cdot \dfrac{5}{4}$

    $\dfrac{10}{3} \cdot x = \dfrac{25}{8}$

    $\dfrac{\frac{10}{3} \cdot x}{\frac{10}{3}} = \dfrac{\frac{25}{8}}{\frac{10}{3}}$

    $x = \dfrac{25}{8} \div \dfrac{10}{3} = \dfrac{\overset{5}{\cancel{25}}}{8} \cdot \dfrac{3}{\underset{2}{\cancel{10}}} = \dfrac{15}{16}$

    Answer: $x = \dfrac{15}{16}$

Section 8.9

3. The left triangle is a right triangle and the right one is not. They are not similar.

7. A corresponds to P
   AC corresponds to PR
   C corresponds to R
   BC corresponds to QR
   B corresponds to Q
   AB corresponds to PQ

11. $\dfrac{AB}{PQ} = \dfrac{9}{6} = \dfrac{3}{2}$

    $\dfrac{AC}{PR} = \dfrac{12}{8} = \dfrac{3}{2}$

    $\dfrac{BC}{QR} = \dfrac{15}{10} = \dfrac{3}{2}$

    All 3 ratios are the same so the triangles are similar triangles.

15. $\dfrac{6}{12} = \dfrac{1}{2}$ Ratio of corresponding sides

    $\dfrac{a}{12} = \dfrac{1}{2}$

    $a = \dfrac{12}{2}$

    $a = 6$

    $\dfrac{b}{15} = \dfrac{1}{2}$

    $b = \dfrac{15}{2}$

    $b = 7.5$

Chapter 8 Review Exercises   145

19. $\frac{2}{16} = \frac{1}{8}$ *Ratio of corresponding sides*

$\frac{3}{n} = \frac{1}{8}$

$1 \cdot n = 3 \cdot 8$

$n = 24$ ft

23.

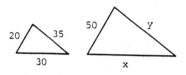

ratio $= \frac{50}{20} = \frac{5}{2}$

$\frac{y}{35} = \frac{5}{2}$

$y = \frac{35 \cdot 2}{2}$

$= 87.5$ m

$\frac{x}{30} = \frac{5}{2}$

$x = \frac{30 \cdot 5}{2}$

$= 75$ m

27. See drawing in textbook.

ratio $= \frac{80}{75} = \frac{16}{15}$

$\frac{m}{80} = \frac{16}{15}$

$m = \frac{80 \cdot 16}{15}$

$m = 85\frac{1}{3} = 85.3$ (rounded)

31. $18 + 7 \cdot 2 - 33 \div 3$

   $= 18 + 14 - 33 \div 3$ *Multiply*

   $= 18 + 14 - 11$ *Divide*

   $= 32 - 11$ *Add*

   $= 21$ *Subtract*

Chapter 8   Review Exercises

1. The figure is a line segment since it is a part of a line and has two endpoints.

2. The figure is a line since it continues forever in opposite directions.

3. The figure is a ray since it is part of a line that continues forever in one direction and has one endpoint.

4. The lines are parallel since they do not intersect.

5. The lines are perpendicular since they intersect and form a right angle.

6. The lines are intersecting, but not perpendicular since they do not form a right angle.

7. ∠TOS is a right angle as indicated by the box in the figure.

8. ∠AOB is an obtuse angle since its measure appears to be more than 90°.

9. ∠P is a straight angle.

10. ∠COD is an acute angle since its measure appears to be less than 90°.

11. Note that ∠1 ≅ ∠4 and ∠2 ≅ ∠5 since they are vertical angles.
    The complementary angles are as follows:
    ∠1 and ∠2;
    ∠4 and ∠5;
    ∠4 and ∠2;
    ∠1 and ∠5.

12. Note that ∠3 ≅ ∠6 and ∠4 ≅ ∠7 since they are vertical angles.
    The complementary angles are as follows:
    ∠1 and ∠2;
    ∠3 and ∠4;
    ∠6 and ∠7;
    ∠3 and ∠7;
    ∠4 and ∠6.

13. ∠AOB and ∠BOC;
    ∠BOC and ∠COD;
    ∠COD and ∠DOA;
    ∠DOA and ∠AOB
    are supplementary since the sum of each pair of angles is 180°.

14. ∠EOH and ∠HOG;
    ∠HOG and ∠GOF;
    ∠GOF and ∠FOE;
    ∠FOE and ∠EOH
    are supplementary since the sum for each pair of angles is 180°.

15. The vertical angles are as follows:
    ∠1 and ∠3;
    ∠2 and ∠4.

16. The vertical angles are as follows:
    ∠1 and ∠3;
    ∠2 and ∠4.

17.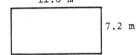

    P = 2l + 2w
    P = 2 · 11.8 + 2 · 7.2
      = 23.6 + 14.4
      = 38 m

18.

    P = 4s
    P = 4 · 42.7
    P = 170.8 cm

19. P = 36 + 24 + 36 + 24
      = 120 m

20. P = 17.9 + 26.5 + 17.9 + 26.5
      = 88.8 m

21. A = lw
      = 27 · 18
      = 486 mm²

22. $A = lw$
    $= 11.7 \cdot 6.42$
    $= 75.114$ or 75.1 ft² (rounded)

23. $A = s^2$
    $= 6\frac{1}{2}^2$ or 6.5^2
    $= 42.25$ or 42.3 m² (rounded)

24. $A = b \cdot h \quad h = 29.7, b = 41.3$
    $= 41.3 \cdot 29.7$
    $= 1226.61$ or 1226.6 cm² (rounded)

25. $A = \frac{1}{2}h(b + B) \quad h = 18, b = 26, B = 37$
    $= \frac{1}{2} \cdot 18(26 + 37)$
    $= \frac{1}{2} \cdot 18 \cdot 63$
    $= 9 \cdot 63$
    $= 567$ ft²

26. $A = \frac{1}{2}h(b + B) \quad h = 31.4, b = 59.7$
    $\qquad\qquad\qquad\qquad B = 72.4$
    $= \frac{1}{2} \cdot 31.4(59.7 + 72.4)$
    $= \frac{1}{2} \cdot 31.4 \cdot 132.1$
    $= 15.7 \cdot 132.1$
    $= 2073.97$ or 2074.0 yd² (rounded)

27. $P = 7 + 12 + 11$
    $= 30$ m

28. $P = 9.4 + 17.2 + 16.8$
    $= 43.4$ cm

29. $A = \frac{1}{2}bh \quad h = 102, b = 212$
    $= \frac{1}{2} \cdot 212 \cdot 102$
    $= 10,812$ cm²

30. $A = \frac{1}{2}bh \quad h = 41.8m \ b = 97.6$
    $= \frac{1}{2} \cdot 97.6 \cdot 41.8$
    $= 2039.84$ ft²

31. $A = \frac{1}{2}bh \quad h = 3\frac{1}{2}, b = 7\frac{1}{4}$
    $= \frac{1}{2} \cdot 7\frac{1}{4} \cdot 3\frac{1}{2}$
    $= \frac{1}{2} \cdot \frac{29}{4} \cdot \frac{7}{2}$
    $= \frac{203}{16} = 12\frac{11}{16}$ ft²

32. $d = 2r \quad r = 72.8$
    $= 2 \cdot 72.8$
    $= 145.6$ m

33. $r = \frac{d}{2} \quad d = 34$
    $= \frac{34}{2}$
    $= 17$ m

34. $C = 2\pi r \quad r = 1$
    $= 2 \cdot 3.14 \cdot 1$
    $= 6.28$ or 6.3 cm (rounded)

    $A = \pi r^2$
    $= 3.14 \cdot 1^2$
    $= 3.14$ or 3.1 cm² (rounded)

35. $C = 2\pi r \qquad r = 17.4$
    $= 2 \cdot 3.14 \cdot 17.4$
    $= 109.272$ or 109.3 in (rounded)

   $A = \pi r^2$
    $= 3.14 \cdot 17.4^2$
    $= 3.14 \cdot 302.76$
    $= 950.6664$ or 950.7 in^2 (rounded)

36. $C = \pi d$
    $= 3.14 \cdot 12$
    $= 37.68$ or 37.7 m (rounded)

   $A = \pi r^2 \qquad r = \dfrac{d}{2} = \dfrac{12}{2} = 6$
    $= 3.14 \cdot 6^2$
    $= 3.14 \cdot 36$
    $= 113.04$ or 113.0 m^2 (rounded)

37. $A = \dfrac{1}{2}\pi r^2 \qquad r = 3.6$
    $= \dfrac{1}{2} \cdot 3.14 \cdot 3.6^2$
    $= \dfrac{1}{2} \cdot 3.14 \cdot 12.96$
    $= 20.3472$ or 20.3 ft^2 (rounded)

38. $V = l \cdot w \cdot h$
    $= 7 \cdot 2 \cdot 5$
    $= 70$ m^3

39. $V = l \cdot w \cdot h$
    $= 9.42 \cdot 3.87 \cdot 2.04$
    $= 74.4$ cm^3 (rounded)

40. $V = l \cdot w \cdot h$
    $= 57 \cdot 174 \cdot 86$
    $= 852{,}948$ mm^3

41. $V = \dfrac{4}{3}\pi r^3$
    $= \dfrac{4}{3} \cdot 3.14 \cdot 5.2^3$
    $= \dfrac{4}{3} \cdot 3.14 \cdot 140.608$
    $= \dfrac{4}{3} \cdot 441.0912$
    $= \dfrac{1766.03648}{3}$
    $= 588.7$ cm^3 (rounded)

42. $A = \dfrac{4}{3}\pi r^3 \qquad r = \dfrac{d}{2} = \dfrac{4.8}{2} = 2.4$
    $= \dfrac{4}{3} \cdot 3.14 \cdot 2.4^3$
    $= \dfrac{4}{3} \cdot 3.14 \cdot 13.824$
    $= \dfrac{173.629}{3}$
    $= 57.9$ cm^3 (rounded)

43. $V = \pi r^2 h$
    $= 3.14 \cdot 3.7^2 \cdot 5.4$
    $= 3.14 \cdot 13.69 \cdot 5.4$
    $= 232.1$ cm^3 (rounded)

44. $V = \pi r^2 h$
    $= 3.14 \cdot 12^2 \cdot 3$
    $= 3.14 \cdot 144 \cdot 3$
    $= 1356.5$ m^3 (rounded)

45. $V = \pi r^2 h$
    $= 3.14 \cdot 50.6^2 \cdot 11.8$
    $= 3.14 \cdot 2560.36 \cdot 11.8$
    $= 94{,}866.5$ in^3 (rounded)

Chapter 8 Review Exercises 149

46. $V = lwh$

    $= 2 \cdot 1\frac{1}{2} \cdot 1\frac{1}{4}$

    $= 2 \cdot \frac{3}{2} \cdot \frac{5}{4}$

    $= \frac{15}{4}$

    $= 3\frac{3}{4}$ m³

47. $V = lwh$

    $= 6 \cdot 4 \cdot 4$

    $= 96$ cm³

48. $V = lwh$

    $= 30 \cdot 20 \cdot 75$

    $= 45,000$ mm³

49. $V = \frac{4}{3}\pi r^3$

    $= \frac{4}{3} \cdot 3.14 \cdot 4^3$

    $= \frac{4}{3} \cdot 3.14 \cdot 64$

    $= \frac{803.84}{3}$

    $= 267.9$ m³ (rounded)

50. $V = \frac{4}{3}\pi r^3$

    $= \frac{4}{3} \cdot 3.14 \cdot 7^3$

    $= \frac{4}{3} \cdot 3.14 \cdot 343$

    $= \frac{4308.08}{3}$

    $= 1436.0$ m³ (rounded)

51. $V = \frac{4}{3}\pi r^3$

    $= \frac{4}{3} \cdot 3.14 \cdot 40^3$

    $= \frac{4}{3} \cdot 3.14 \cdot 64,000$

    $= 267,946.7$ mm³ (rounded)

52. $V = \pi r^2 h$

    $= 3.14 \cdot 5^2 \cdot 7$

    $= 3.14 \cdot 25 \cdot 7$

    $= 549.5$ cm³

53. $V = \pi r^2 h$

    $= 3.14 \cdot 12^2 \cdot 4$

    $= 3.14 \cdot 144 \cdot 4$

    $= 1808.6$ m³ (rounded)

54. $V = \pi r^2 h$

    $= 3.14 \cdot 5^2 \cdot 32$

    $= 3.14 \cdot 25 \cdot 32$

    $= 2512$ cm³

55. $\sqrt{7} = 2.646$

56. $\sqrt{19} = 4.359$

57. $\sqrt{27} = 5.196$

58. $\sqrt{35} = 5.916$

59. $\sqrt{58} = 7.616$

60. $\sqrt{121} = 11$

61. $\sqrt{144} = 12$

62. $\sqrt{169} = 13$

63. Hypotenuse $= \sqrt{8^2 + 15^2}$
    $= \sqrt{64 + 225}$
    $= \sqrt{289}$
    $= 17$ in

64. leg $= \sqrt{25^2 - 24^2}$
    $= \sqrt{625 - 576}$
    $= \sqrt{49}$
    $= 7$ cm

65. leg $= \sqrt{15^2 - 11^2}$
    $= \sqrt{225 - 121}$
    $= \sqrt{104}$
    $= 10.198$ cm

66. Hypotenuse $= \sqrt{6^2 + 4^2}$
    $= \sqrt{36 + 16}$
    $= \sqrt{52}$
    $= 7.211$ in

67. leg $= \sqrt{13^2 - 9^2}$
    $= \sqrt{169 - 81}$
    $= \sqrt{88}$
    $= 9.381$ cm

68. leg $= \sqrt{15^2 - 9^2}$
    $= \sqrt{225 - 81}$
    $= \sqrt{144}$
    $= 12$ cm

69. $\frac{20}{40} = \frac{1}{2}$ Ratio of corresponding sides
    $\frac{15}{y} = \frac{1}{2}$
    $y = 15 \cdot 2 = 30$
    $\frac{17}{x} = \frac{1}{2}$
    $x = 17 \cdot 2 = 34$

70. $\frac{12}{8} = \frac{3}{2}$ Ratio of corresponding sides
    $\frac{x}{12} = \frac{3}{2}$
    $x = \frac{12 \cdot 3}{2} = 18$
    $\frac{y}{10} = \frac{3}{2}$
    $y = \frac{10 \cdot 3}{2} = 15$

71. $\frac{15}{10} = \frac{3}{2}$ Ratio of corresponding sides
    $\frac{m}{9} = \frac{3}{2}$
    $m = \frac{9 \cdot 3}{2} = 13\frac{1}{2}$ or 13.5
    $\frac{n}{8} = \frac{3}{2}$
    $n = \frac{8 \cdot 3}{2} = 12$

72.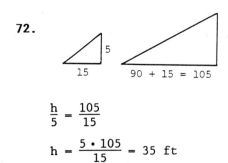

    $\frac{h}{5} = \frac{105}{15}$
    $h = \frac{5 \cdot 105}{15} = 35$ ft

73.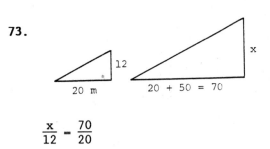

    $\frac{x}{12} = \frac{70}{20}$
    $x = \frac{12 \cdot 70}{20} = 42$ m

74.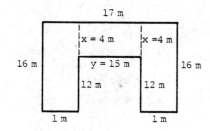

Note that x = 16 - 12 = 4 m and
y = 17 - (1 + 1)
  = 17 - 2 = 15 m.
P = 16 + 17 + 16 + 1 + 12 + 15 + 12 + 1
  = 90 m
A (left)   = 16 · 1 = 16 m²
A (middle) = 4 · 15 = 60 m²
A (right) = 16 · 1 = 16 m²
A (total) = 16 + 60 + 16 = 92 m²

75. P = 50 + 72 + 19 + 50 + 19 + 72
     = 282 cm
View the figure as a left parallelogram and a right parallelogram.
A (left) = 45 · 72 = 3240
A (right) = 19 · 50 = 950
A (total) = 3240 + 950 = 4190 cm²

76. $V = \pi r^2 h$
   = 3.14 · 2² · 8
   = 3.14 · 4 · 8
   = 100.5 ft³ (rounded)

77. $V = \pi r^2 h$
   = 3.14 · 1.5² · 1.5
   = 3.14 · 2.25 · 1.5
   = 10.6 in³ (rounded)

78. $V = \frac{1}{3} B \cdot h$ B = 9 · 11 = 99
   = $\frac{1}{3}$ · 99 · 17
   = 561 cm³

79. $V = \frac{1}{3} \pi r^2 h$
   = $\frac{1}{3}$ · 3.14 · 9² · 15
   = $\frac{1}{3}$ · 3.14 · 81 · 15
   = 1271.7 cm³

80. $V = \frac{1}{3} \pi r^2 h$
   = $\frac{1}{3}$ · 3.14 · 7² · 25
   = $\frac{1}{3}$ · 3.14 · 49 · 25
   = 1282.2 cm³ (rounded)

81. V = Bh
   = 25 · 12
   = 300 cm³

82. Assume that the top triangle, DEC, and the entire triangle, ABC, are similar triangles.
Sides CD and CA are corresponding sides. The ratio of their lengths is

$$\frac{3}{3+2} \text{ or } \frac{3}{5}.$$

Since the triangles are similar, the ratio of any corresponding sides will also equal $\frac{3}{5}$. Since DE and AB are corresponding sides, $\frac{DE}{AB} = \frac{3}{5}$.

152 Chapter 8 Geometry

Replace DE by 4 and AB by x to obtain

$$\frac{4}{x} = \frac{3}{5}$$

$$3 \cdot x = 4 \cdot 5$$

$$\frac{3 \cdot x}{3} = \frac{20}{3}$$

$$x = \frac{20}{3}$$

Answer: $AB = \frac{20}{3}$

83. Split the solid into two pieces as shown.

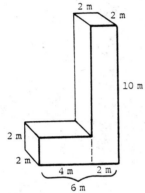

$V\text{(left)} = 2 \cdot 4 \cdot 2$
$= 16 \text{ m}^3$

$V\text{(right)} = 2 \cdot 2 \cdot 10$
$= 40 \text{ m}^3$

$V\text{(total)} = 16 + 40$
$= 56 \text{ m}^3$

84. If each side of a cube had length s, then its volume would be

$V = lwh = s \cdot s \cdot s = s^3.$

If each side were then doubled in length to 2s, then the new volume would be

$V = lwh = 2s \cdot 2s \cdot 2s = 8s^3.$

Note that the volume of the second cube is 8 times the volume of the first cube.

Answer: 8

85. $A\text{(rectangle)} = lw$

$w = 21, \; l = 15 + 15 + 15 = 45$

$= 45 \cdot 21$

$= 945 \text{ m}^2$

$A\text{(triangle)} = \frac{1}{2}bh \quad b = 15, \; h = 10$

$= \frac{1}{2} \cdot 15 \cdot 10$

$= 75 \text{ m}^2$

$A\text{(total)} = 945 + 75$

$= 1020 \text{ m}^2$

86. $A\text{(large rectangle)} = l \cdot w$

$w = 15, \; l = 6 + 7 + 6 = 19$

$= 19 \cdot 15$

$= 285 \text{ ft}^2$

$A\text{(small rectangle)} = l \cdot w$

$l = 8, \; w = 7$

$= 8 \cdot 7$

$= 56 \text{ ft}^2$

$A\text{(shaded)} = 285 - 56$

$= 229 \text{ ft}^2$

87. $A\text{(trapezoid)} = \frac{1}{2}h(b + B)$

$h = 12, \; b = 11, \; B = 18$

$= \frac{1}{2} \cdot 12(11 + 18)$

$= \frac{1}{2} \cdot 12 \cdot 29$

$= 174 \text{ ft}^2$

$$A \text{ (triangle)} = \tfrac{1}{2}bh \quad b = 12, h = 7$$
$$= \tfrac{1}{2} \cdot 12 \cdot 7$$
$$= 42 \text{ ft}^2$$
$$A \text{ (shaded)} = 174 - 42$$
$$= 132 \text{ ft}^2$$

88. $A \text{ (rectangle)} = lw$
$l = 48 + 48 = 96, w = 74$
$$= 96 \cdot 74$$
$$= 7104 \text{ cm}^2$$
$$A \text{ (triangle)} = \tfrac{1}{2}bh \quad b = 48, h = 36$$
$$= \tfrac{1}{2} \cdot 48 \cdot 36$$
$$= 864 \text{ cm}^2$$
$A \text{ (both triangles)} = 864 \cdot 2$
(They have the same height and base)
$$= 1728 \text{ cm}^2$$
$A \text{ (shaded)} = 7104 - 1728$
$$= 5376 \text{ cm}^2$$

89. $A \text{ (rectangle)} = lw \quad l = 32, w = 21$
$$= 32 \cdot 21$$
$$= 672 \text{ ft}^2$$
$$A \text{ (semicircle)} = \tfrac{1}{2}\pi r^2$$
$$r = \tfrac{d}{2} = \tfrac{21}{2} = 10.5$$
$$= \tfrac{1}{2} \cdot 3.14 \cdot 10.5^2$$
$$= \tfrac{1}{2} \cdot 3.14 \cdot 110.25$$
$$= 173.0925 \text{ ft}^2$$
$A \text{ (shaded)} = 672 - 173.0925$
$$= 498.9075$$
$$= 498.9 \text{ ft}^2 \text{ (rounded)}$$

90. $A \text{ (rectangle)} = lw \quad l = 21, w = 14$
$$= 21 \cdot 14$$
$$= 294 \text{ cm}^2$$
The two end semicircles make one circle.
$$A \text{ (circle)} = \pi r^2 \quad r = 7$$
$$= 3.14 \cdot 7^2$$
$$= 3.14 \cdot 49$$
$$= 153.86 \text{ cm}^2$$
$A \text{ (shaded)} = 294 + 153.86$
$$= 447.86$$
$$= 447.9 \text{ cm}^2 \text{ (rounded)}$$

Chapter 8 Test

1. The figure is a ray since it is part of a line that has one endpoint and continues forever in one direction.

2. The figure is a line since it continues forever in opposite directions.

3. ∠POQ is an acute angle since its measure is less than 90°.

4. ∠AOB is a right angle since its measure is 90°.

5. The complement of a 72° angle is 18° since 90° − 72° = 18°.

6. The supplement of a 110° angle is 70° since 180° − 110° = 70°.

7. The vertical angles are as follows:
    ∠1 and ∠4;
    ∠2 and ∠5;
    ∠3 and ∠6.

8. The congruent angles are as follows:
    ∠1 and ∠4;
    ∠2 and ∠5;
    ∠3 and ∠6.

9. $P = 2l + 2w$
    $= 2 \cdot 46.8 + 2 \cdot 39.4$
    $= 93.6 + 78.8$
    $= 172.4$ cm

   $A = lw$
    $= 46.8 \cdot 39.4$
    $= 1843.92$ cm²

10. $P = 4s$
    $= 4 \cdot 14.9$
    $= 59.6$ m

    $A = s^2$
    $= 14.9^2$
    $= 222.01$ m²

11. $A = bh$
    $= 72 \cdot 46$
    $= 3312$ m²

12. $A = \frac{1}{2}h(b + B)$
    $= \frac{1}{2} \cdot 37(29 + 57)$
    $= \frac{1}{2} \cdot 37 \cdot 86$
    $= 1591$ cm²

13. $A = \frac{1}{2}bh$
    $= \frac{1}{2} \cdot 12 \cdot 8$
    $= 48$ m²

14. $A = \frac{1}{2}bh$
    $= \frac{1}{2} \cdot 11.8 \cdot 15.4$
    $= 90.86$ cm²

15. $P = 51.4 + 29.7 + 38.2$
    $= 119.3$ m

16. $A = \pi r^2$
    $= 3.14 \cdot 12^2$
    $= 3.14 \cdot 144$
    $= 452.2$ m² (rounded)

17. $A = \pi r^2 \quad r = \frac{d}{2} = \frac{39.4}{2} = 19.7$
    $= 3.14 \cdot 19.7^2$
    $= 3.14 \cdot 388.09$
    $= 1218.6$ cm² (rounded)

18. $A = \frac{1}{2}\pi r^2$
    $= \frac{1}{2} \cdot 3.14 \cdot 6^2$
    $= \frac{1}{2} \cdot 3.14 \cdot 36$
    $= 56.5$ m² (rounded)

19. $C = \pi d$
    $= 3.14 \cdot 25$
    $= 78.5$ in

20. $V = lwh$
    $= 30 \cdot 18 \cdot 12$
    $= 6480 \text{ m}^3$

21. $V = \frac{4}{3}\pi r^3$
    $= \frac{4}{3} \cdot 3.14 \cdot 9^3$
    $= \frac{4}{3} \cdot 3.14 \cdot 729$
    $= 3052.08 \text{ m}^3$

22. $V = \frac{1}{3}\pi r^2 h$
    $= \frac{1}{3} \cdot 3.14 \cdot 6^2 \cdot 8$
    $= \frac{1}{3} \cdot 3.14 \cdot 36 \cdot 8$
    $= 301.44 \text{ m}^3$

23. $\text{leg} = \sqrt{15^2 - 9^2}$
    $= \sqrt{225 - 81}$
    $= \sqrt{144}$
    $= 12 \text{ cm}$

24. $\text{hypotenuse} = \sqrt{6^2 + 7^2}$
    $= \sqrt{36 + 49}$
    $= \sqrt{85}$
    $= 9.220 \text{ cm}$

25. $\frac{4}{8} = \frac{1}{2}$ *Ratio of corresponding sides*
    $\frac{5}{q} = \frac{1}{2}$
    $q = 5 \cdot 2 = 10$
    $\frac{6}{p} = \frac{1}{2}$
    $p = 6 \cdot 2 = 12$

26. $\frac{10}{15} = \frac{2}{3}$ *Ratio of corresponding sides*
    $\frac{y}{18} = \frac{2}{3}$
    $y = \frac{18 \cdot 2}{3} = 12$
    $\frac{z}{9} = \frac{2}{3}$
    $z = \frac{9 \cdot 2}{3} = 6$

CHAPTER 9 BASIC ALGEBRA
Section 9.1

3. −12

7. positive

11. neither

15. See graph in the answer section of the textbook.

19. See graph in the answer section of the textbook.

23. See graph in the answer section of the textbook.

27. $5 > 4$
Because 5 is to the right of 4 on a number line.

31. $8 > -6$
Because 8 is to the right of −6 on a number line.

35. $-8 < -3$
Because −8 is to the left of −3 on a number line.

39. $0 > -2$
Because 0 is to the right of −2 on a number line.

43. $|-5| = 5$ (−5 is 5 units from 0)

47. $|251| = 251$ (251 is 251 units from 0)

51. $\left|-\frac{1}{2}\right| = \frac{1}{2}$ ($-\frac{1}{2}$ is $\frac{1}{2}$ a unit from 0)

55. $|8.3| = 8.3$ (8.3 is 8.3 units from 0)

59. $\left|-\frac{9}{7}\right| = \frac{9}{7}$ ($-\frac{9}{7}$ is $\frac{9}{7}$ units from 0)

63. $-|4| = -(4) = -4$
(4 is 4 units from 0, then take its opposite)

67. $-(-5) = 5$

71. $-(16) = -16$

75. $-\left(-\frac{1}{2}\right) = \frac{1}{2}$

79. $-(-1.4) = 1.4$

83. True
Since −8 is to the left of −4 on a number line.

87. True
because $-|-7| = -7$ and $-(3) = -3$; $-7 < -3$ is true since −7 is to the left of −3 on a number line.

91. $\begin{aligned} 2 &= 2 \\ 1 &= 1 \\ \tfrac{2}{5} &= \tfrac{4}{10} \\ +\,1\tfrac{1}{10} &= 1\tfrac{1}{10} \\ \hline 4\tfrac{5}{10} &= 4\tfrac{1}{2} \end{aligned}$

Section 9.2

3. −5 + (−2) = −7
See graph in the answer section of the textbook.

7. −8 + 5   Different signs, so subtract absolute value: $|-8| - |5| = 8 - 5 = 3$. Use sign of larger absolute value (−8). Answer is −3.

11. −2 + (−5)   Same signs, so add absolute values:
$|-2| + |-5| = 2 + 5 = 7$.
Keep sign. Answer is −7.

15. −8 + (−9)   Same signs, so add absolute values:
$|-8| + |-9| = 8 + 9 = 17$.
Keep sign. Answer is −17.

19. 7.8 + (−14.6)   Different signs, so subtract absolute values:
$|-14.6| - |7.8| = 14.6 - 7.8 = 6.8$.
Use sign of larger absolute value (−14.6). Answer is −6.8.

23. $-\frac{5}{8} + \frac{1}{4}$ Different signs so subtract absolute values:
$|-\frac{5}{8}| - |\frac{1}{4}| = \frac{5}{8} - \frac{1}{4} = \frac{5}{8} - \frac{2}{8} = \frac{3}{8}$
Use the sign of the larger absolute value. Answer is $-\frac{3}{8}$.

27. $\begin{array}{r} -9 \\ -2 \\ \hline -11 \end{array}$ Add the absolute values and keep sign.

31. The additive inverse of 3 is −3. (change sign)

35. The additive inverse of −8 is 8. (change sign)

39. The additive inverse of $\frac{1}{4}$ is $-\frac{1}{4}$. (change sign)

43. 6 − 9 = 6 + (−9) = −3

47. 5 − 15 = 5 + (−15) = −10

51. −2 − 15 = −2 + (−15) = −17

55. 7 − (−9) = 7 + 9 = 16

59. −4 − (−9) = −4 + 9 = 5

63. $\left.\begin{array}{r}-6 \\ -2\end{array}\right]$ $\begin{array}{r}-8\\ 5\\\hline -3\end{array}$
   5

67. $\left.\begin{array}{r}-9 \\ -8\end{array}\right]$ −17
   $\left.\begin{array}{r}2 \\ 5\end{array}\right]$ 13
   $\begin{array}{r}6 \\ \hline -4\end{array}$

71. 4 − (−13) + (−5)
    = 4 + 13 + (−5)
    = 17 + (−5)
    = 12

75. $\frac{1}{2} - \frac{2}{3} + (-\frac{5}{6}) = \frac{1}{2} + (-\frac{2}{3}) + (-\frac{5}{6})$
    $= -\frac{1}{6} + (-\frac{5}{6})$
    $= -\frac{6}{6} = -1$

79. −2 + (−11) + |−2|
    = −13 + |−2|
    = −13 + 2
    = −11

83. $2\frac{1}{2} + 3\frac{1}{4} - (-1\frac{3}{8}) - 2\frac{3}{8}$
    $= 2\frac{4}{8} + 3\frac{2}{8} - (-1\frac{3}{8}) - 2\frac{3}{8}$
    $= 5\frac{6}{8} - (-1\frac{3}{8}) - 2\frac{3}{8}$
    $= 5\frac{6}{8} + 1\frac{3}{8} - 2\frac{3}{8}$
    $= 6\frac{9}{8} - 2\frac{3}{8}$
    $= 4\frac{6}{8} = 4\frac{3}{4}$

87. $\frac{8}{11} \cdot \frac{3}{5} = \frac{8 \cdot 3}{11 \cdot 5} = \frac{24}{55}$

91. $1\frac{2}{3} \div 2\frac{7}{9}$
    $= \frac{5}{3} \div \frac{25}{9}$
    $= \frac{\cancel{5}^1}{\cancel{3}_1} \cdot \frac{\cancel{9}^3}{\cancel{25}_5}$
    $= \frac{1 \cdot 3}{1 \cdot 5} = \frac{3}{5}$

Section 9.3

3. −9 · 3 = −27 (Different signs, product is negative.)

7. 10 · (−5) = −50 (Different signs, product is negative.)

11. −8 · (−4) = 32 (Same signs, product is positive.)

15. −19 · (−7) = 133 (Same signs, product is positive.)

19. $-\frac{3}{4} \cdot 12 = -\frac{3}{\cancel{4}_1} \cdot \frac{\cancel{12}^3}{1} = -3 \cdot 3 = -9$

23. $-10 \cdot (\frac{2}{5}) = -\frac{\cancel{10}^2}{1} \cdot \frac{2}{\cancel{5}_1} = -2 \cdot 2 = -4$

27. $-\frac{7}{\cancel{8}_1} \cdot \frac{\cancel{10}^2}{3} = -\frac{7 \cdot 2}{3} = -\frac{14}{3}$ or $-4\frac{2}{3}$

31. $-\frac{7}{\cancel{4}_\cancel{2}} \cdot (-\frac{\cancel{8}^2}{3}) = \frac{-7 \cdot (-2)}{3} = \frac{14}{3} = 4\frac{2}{3}$

35. −11.4 · (−18) = 205.2

39. −11.2 · (−4.2) = 47.04

43. $\frac{-25}{5} = -5$ (Different signs, quotient is negative.)

9.4 Order of Operations   159

47. $\dfrac{-86}{2} = -43$ (Different signs, quotient is negative.)

51. $\dfrac{-72}{-36} = 2$ (Same signs, quotient is positive.)

55. $-\dfrac{1}{9} + \left(-\dfrac{5}{18}\right)$

$= -\dfrac{1}{\cancel{9}_1} \cdot \left(-\dfrac{\cancel{18}^2}{5}\right) = \dfrac{-1 \cdot (-2)}{1 \cdot 5} = \dfrac{2}{5}$

59. $-\dfrac{2}{3} \div (-2)$

$= -\dfrac{\cancel{2}^1}{3} \cdot \left(-\dfrac{1}{\cancel{2}_1}\right) = \dfrac{1}{3}$

63. $\dfrac{-10.72}{2} = -5.36$

67. $\dfrac{-9.75}{2.5} = -3.9$

71. $(-4) \cdot (-6) \cdot \dfrac{1}{2}$

$= 24 \cdot \dfrac{1}{2}$

$= 12$

75. $\left(-\dfrac{1}{\cancel{2}_1}\right) \cdot \left(-\dfrac{\cancel{2}^1}{5}\right) \cdot \left(\dfrac{7}{8}\right)$

$= \left(\dfrac{1}{5}\right) \cdot \left(\dfrac{7}{8}\right)$

$= \dfrac{7}{40}$

79. $(-12) \div (-2) \div (4 \div 2)$
$= (-12) \div (-2) \div (2)$
$= 6 \div (2)$
$= 3$

83. $7 + 6 \div 2 \cdot 4 - 9$
$= 7 + 3 \cdot 4 - 9$
$= 7 + 12 - 9$
$= 19 - 9$
$= 10$

Section 9.4

3. $-8 + 7 + (-9) \cdot 2$
$= -8 + 7 + (-18)$
$= -1 + (-18)$
$= -19$

7. $9 - 5^2$
$= 9 - 25$
$= -16$

11. $4^2 + 3^2 + (-8)$
$= 16 + 9 + (-8)$
$= 25 + (-8)$
$= 17$

15. $(-4)^2 + (-3)^2 + 5$
$= 16 + 9 + 5$
$= 30$

19. $-7 + 6 \cdot (8 - 14)$
$= -7 + 6 \cdot (-6)$
$= -7 + (-36)$
$= -43$

23. $(-5) \cdot (7 - 13) \div (-10)$
$= (-5) \cdot (-6) \div (-10)$
$= 30 \div (-10)$
$= -3$

27. $2 - (-5) \cdot (-3)^2$
$= 2 - (-5) \cdot (9)$
$= 2 - (-45)$
$= 47$

31. $36 \div (-4) - 25 \div (-5)$
$= -9 - (-5)$
$= -9 + (5)$
$= -4$

35. $4 \cdot 3^2 + 7 \cdot (3 + 9) - (-6)$
$= 4 \cdot 9 + 7 \cdot (12) + 6$
$= 36 + 84 + 6$
$= 126$

39. $2 \cdot 3^2 - 4 \cdot (7 - 3) - 5^2$
$= 2 \cdot 9 - 4 \cdot (4) - 25$
$= 18 - 16 - 25$
$= 2 - 25$
$= -23$

43. $(-4)^2 \cdot (7 - 9)^2 \div 2^3$
$= 16 \cdot (-2)^2 \div 8$
$= 16 \cdot 4 \div 8$
$= 64 \div 8$
$= 8$

47. $(-2.1) \cdot (1.8 - 2.7)^2$
$= (-2.1) \cdot (-.9)^2$
$= (-2.1) \cdot (.81)$
$= -1.701$

51. $\frac{2}{3} \div \left(-\frac{5}{6}\right) - \frac{1}{2}$

$= \frac{2}{\overset{}{\underset{1}{\cancel{3}}}} \cdot \left(-\frac{\overset{2}{\cancel{6}}}{5}\right) - \frac{1}{2}$

$= \frac{2 \cdot (-2)}{5} - \frac{1}{2}$

$= -\frac{4}{5} + \left(-\frac{1}{2}\right)$

$= -\frac{8}{10} + \left(-\frac{5}{10}\right)$

$= -\frac{13}{10}$ or $-1\frac{3}{10}$

55. $\frac{\overset{1}{\cancel{3}}}{5} \cdot \left(-\frac{7}{\underset{2}{\cancel{6}}}\right) - \left(\frac{1}{6} - \frac{5}{3}\right)$

$= -\frac{7}{10} - \left(\frac{1}{6} - \frac{10}{6}\right)$

$= -\frac{7}{10} - \left(-\frac{9}{6}\right)$

$= -\frac{7}{10} + \frac{9}{6}$

$= -\frac{7}{10} + \frac{3}{2}$

$= -\frac{7}{10} + \frac{15}{10}$

$= \frac{8}{10}$

$= \frac{4}{5}$

59. $4.2 \cdot (-1.6) \div (-.56) \div 2^2$
$= -6.72 \div (-.56) \div 4$
$= 12 \div 4$
$= 3$

63. $\frac{1}{3} \cdot (-\frac{1}{2}) + (-\frac{\overset{1}{\cancel{5}}}{8}) \cdot (-\frac{7}{\underset{2}{\cancel{10}}})$

$= -\frac{1}{6} + (\frac{7}{16})$

$= -\frac{8}{48} + (\frac{21}{48})$

$= \frac{13}{48}$

67. $-7 \cdot (6 - \frac{5}{8} \cdot \overset{3}{\underset{1}{\cancel{24}}} + 3 \cdot \frac{8}{3})$

$= -7 \cdot (6 - 15 + 1 \cdot 8)$

$= -7 \cdot (6 - 15 + 8)$

$= -7 \cdot (-9 + 8)$

$= -7 \cdot (-1)$

$= 7$

71. $\frac{\overset{2}{\cancel{6}}}{\underset{1}{\cancel{7}}} \cdot \frac{\overset{2}{\cancel{14}}}{\underset{3}{\cancel{9}}} = \frac{2 \cdot 2}{1 \cdot 3} = \frac{4}{3}$ or $1\frac{1}{3}$

Section 9.5

3. $r = 5, \; s = -3$
    $2r + 4s = 2(5) + 4(-3)$
    $= 10 + (-12)$
    $= -2$

7. $r = -2, \; s = -1$
    $2r + 4s = 2(-2) + 4(-1)$
    $= -4 + (-4)$
    $= -8$

11. $8x - y; \; x = 1, \; y = 2$
    $= 8(1) - (2)$
    $= 8 - 2$
    $= 6$

15. $\frac{-m + 7n}{s + 1}; \; m = 2, \; n = -1, \; s = 2$
    $= \frac{-(2) + 7(-1)}{2 + 1}$
    $= \frac{-2 + (-7)}{3}$
    $= \frac{-9}{3}$
    $= -3$

19. $P = 4s; \; s = 7$
    $= 4 \cdot (7)$
    $= 28$

23. $A = \frac{1}{2}bh; \; b = 15, \; h = 20$
    $= \frac{1}{2} \cdot (15) \cdot (20)$
    $= 150$

27. $V = \frac{1}{3}Bh; \; B = 40, \; h = 6$
    $= \frac{1}{3} \cdot (40) \cdot (6)$
    $= 80$

31. $C = 2\pi r; \; \pi = 3.14, \; r = 7$
    $= 2 \cdot (3.14) \cdot (7)$
    $= 43.96$

35. C = -40

$$F = \frac{9}{5}C + 32$$

$$= \frac{9}{5} \cdot (-40) + 32$$

$$= \frac{9}{\overset{1}{5}} \cdot \frac{\overset{-8}{-40}}{1} + 32$$

$$= -72 + 32$$

$$= -40$$

39. 14 - 21 + 7

= -7 + 7

= 0

Section 9.6

3. 3y = 27; 9

3(9) = 27

27 = 27

Solution is 9, so the answer is yes.

7. p + 5 = 9

p + 5 - 5 = 9 - 5   *Subtract 5 from each side*

p = 4

Check:

p + 5 = 9

4 + 5 = 9

9 = 9   *True*

11. z - 5 = 3

z - 5 + 5 = 3 + 5   *Add the inverse of -5 to each side*

z = 8

Check:

z - 5 = 3

8 - 5 = 3

3 = 3   *True*

15. -4 = n + 2

-4 - 2 = n + 2 - 2   *Subtract 2 from each side*

-6 = n

Check:

-4 = n + 2

-4 = -6 + 2

-4 = -4   *True*

19. k + 15 = 2

k + 15 - 15 = 2 - 15

k = -13

Check:

k + 15 = 2

-13 + 15 = 2

2 = 2   *True*

23. -7 + r = -8

-7 + 7 + r = -8 + 7

r = -1

Check:

-7 + r = -8

-7 + (-1) = -8

-8 = -8   *True*

27. $z + \frac{5}{8} = 2$

$z + \frac{5}{8} - \frac{5}{8} = 2 - \frac{5}{8}$

$z = \frac{16}{8} - \frac{5}{8}$

$= \frac{11}{8} = 1\frac{3}{8}$

Check:

$$z + \frac{5}{8} = 2$$

$$1\frac{3}{8} + \frac{5}{8} = 2$$

$$1\frac{8}{8} = 2$$

$$1 + \frac{8}{8} = 2$$

$$1 + 1 = 2$$

$$2 = 2 \quad True$$

31.
$$m - 1\frac{4}{5} = 2\frac{1}{10}$$

$$m - 1\frac{4}{5} + 1\frac{4}{5} = 2\frac{1}{10} + 1\frac{4}{5}$$

$$m + 0 = 2\frac{1}{10} + 1\frac{8}{10}$$

$$m = 3\frac{9}{10}$$

Check:

$$3\frac{9}{10} - 1\frac{4}{5} = 2\frac{1}{10}$$

$$3\frac{9}{10} - 1\frac{8}{10} = 2\frac{1}{10}$$

$$2\frac{1}{10} = 2\frac{1}{10} \quad True$$

35.
$$4.76 + r = 3.25$$

$$4.76 - 4.76 + r = 3.25 - 4.76$$

$$0 + r = -1.51$$

$$r = -1.51$$

Check:

$$4.76 + (-1.51) = 3.25$$

$$3.25 = 3.25 \quad True$$

39. $12r = 48$

$$\frac{12r}{12} = \frac{48}{12} \quad \text{Divide both sides by 12}$$

$$r = 4$$

Check:

$$12r = 48$$

$$12 \cdot 4 = 48$$

$$48 = 48 \quad True$$

43. $-6k = 36$

$$\frac{-6k}{-6} = \frac{36}{-6} \quad \text{Divide both sides by -6}$$

$$k = -6$$

Check:

$$-6k = 36$$

$$-6(-6) = 36$$

$$36 = 36 \quad True$$

47. $-1.2m = 8.4$

$$\frac{-1.2m}{-1.2} = \frac{8.4}{-1.2} \quad \text{Divide both sides by -1.2}$$

$$m = -7$$

Check:

$$-1.2m = 8.4$$

$$-1.2(-7) = 8.4$$

$$8.4 = 8.4 \quad True$$

51. $\frac{k}{2} = 17$

$$2 \cdot \frac{k}{2} = 17 \cdot 2 \quad \text{Multiply each side by 2}$$

$$k = 34$$

Check:

$$\frac{k}{2} = 17$$

$$\frac{34}{2} = 17$$

$$17 = 17 \quad True$$

55. $\dfrac{r}{3} = -12$

$3 \cdot \dfrac{r}{3} = -12 \cdot 3$

$r = -36$

Check:

$\dfrac{r}{3} = -12$

$-\dfrac{36}{3} = -12$

$-12 = -12$ *True*

59. $-\dfrac{1}{4}m = 2$

$(-4)\left(-\dfrac{1}{4}m\right) = 2 \cdot (-4)$

$m = -8$

Check:

$-\dfrac{1}{4}m = 2$

$-\dfrac{1}{4}(-8) = 2$

$2 = 2$ *True*

63. $-\dfrac{4}{7}y = 16$

$\left(-\dfrac{7}{4}\right) \cdot \left(-\dfrac{4}{7}y\right) = 16 \cdot \left(-\dfrac{7}{4}\right)$

$y = -28$

Check:

$-\dfrac{4}{7}y = 16$

$-\dfrac{4}{7}(-28) = 16$

$-\dfrac{4}{\cancel{7}} \cdot \dfrac{\overset{-4}{\cancel{-28}}}{1} = 16$

$16 = 16$ *True*

67. $\dfrac{y}{1.7} = .8$

$(1.7)\dfrac{y}{1.7} = .8(1.7)$

$y = 1.36$

Check:

$\dfrac{y}{1.7} = .8$

$\dfrac{1.36}{1.7} = .8$

$.8 = .8$ *True*

71. $x - 17 = 5 - 3$

$x - 17 = 2$

$x - 17 + 17 = 2 + 17$

$x + 0 = 19$

$x = 19$

75. $\dfrac{7}{2}x = \dfrac{4}{3}$

$\dfrac{2}{7} \cdot \left(\dfrac{7}{2}x\right) = \dfrac{2}{7}\left(\dfrac{4}{3}\right)$

$x = \dfrac{8}{21}$

79. $\dfrac{1}{2} + \dfrac{\overset{1}{\cancel{2}}}{\cancel{4}} \cdot \dfrac{\overset{2}{\cancel{8}}}{\cancel{9}} - \dfrac{1}{6}$

$= \dfrac{1}{2} + \dfrac{2}{3} - \dfrac{1}{6}$

$= \dfrac{3}{6} + \dfrac{4}{6} - \dfrac{1}{6}$

$= \dfrac{7}{6} - \dfrac{1}{6}$

$= \dfrac{6}{6}$ or 1

Section 9.7

3. $2y - 8 = 14$
    $2y - 8 + 8 = 14 + 8$
    $\dfrac{2y}{y} = \dfrac{22}{2}$
    $y = 11$

    Check:
    $2y - 8 = 14$
    $2 \cdot 11 - 8 = 14$
    $22 - 8 = 14$
    $14 = 14$ *True*

7. $-8a + 7 = 23$
    $-8a + 7 - 7 = 23 - 7$
    $\dfrac{-8a}{-8} = \dfrac{16}{-8}$
    $a = -2$

    Check:
    $-8a + 7 = 23$
    $-8(-2) + 7 = 23$
    $16 + 7 = 23$
    $23 = 23$ *True*

11. $-\tfrac{1}{2}z + 2 = -1$
    $-\tfrac{1}{2}z + 2 - 2 = -1 - 2$
    $-\tfrac{1}{2}z = -3$
    $-\tfrac{2}{1} \cdot (-\tfrac{1}{2}z) = -3 \cdot (-\tfrac{2}{1})$
    $z = 6$

    Check:
    $-\tfrac{1}{2}z + 2 = -1$
    $-\tfrac{1}{2}(6) + 2 = -1$
    $-3 + 2 = -1$
    $-1 = -1$ *True*

15. $5(r + 3)$
    $= 5 \cdot r + 5 \cdot 3$
    $= 5r + 15$

19. $-2(y - 3)$
    $= -2y + (-2) \cdot (-3)$
    $= -2y + 6$

23. $2m + 5m$
    $= (2 + 5)m$
    $= 7m$

27. $9z - 4z$
    $= (9 - 4)z$
    $= 5z$

31. $4k + 6k = 50$
    $(4 + 6)k = 50$
    $10k = 50$
    $\dfrac{10k}{10} = \dfrac{50}{10}$
    $k = 5$

    Check:
    $4k + 6k = 50$
    $4 \cdot 5 + 6 \cdot 5 = 50$
    $20 + 30 = 50$
    $50 = 50$ *True*

35.
$$2b - 6b = 24$$
$$(2 - 6)b = 24$$
$$-4b = 24$$
$$\frac{-4b}{-4} = \frac{24}{-4}$$
$$b = -6$$

Check:
$$2b - 6b = 24$$
$$2(-6) - 6(-6) = 24$$
$$-12 - (-36) = 24$$
$$-12 + 36 = 24$$
$$24 = 24 \quad True$$

39.
$$6p - 2 = 4p + 6$$
$$6p - 2 + 2 = 4p + 6 + 2$$
$$6p = 4p + 8$$
$$6p - 4p = 4p - 4p + 8$$
$$2p = 8$$
$$\frac{2p}{2} = \frac{8}{2}$$
$$p = 4$$

Check:
$$6p - 2 = 4p + 6$$
$$6 \cdot 4 - 2 = 4 \cdot 4 + 6$$
$$24 - 2 = 16 + 6$$
$$22 = 22 \quad True$$

43.
$$2.5y + 6 = 4.5y + 10$$
$$2.5y + 6 - 6 = 4.5y + 10 - 6$$
$$2.5y = 4.5y + 4$$
$$2.5y - 4.5y = 4.5y - 4.5y + 4$$
$$-2y = 4$$
$$\frac{-2y}{-2} = \frac{4}{-2}$$
$$y = -2$$

Check:
$$2.5y + 6 = 4.5y + 10$$
$$2.5(-2) + 6 = 4.5(-2) + 10$$
$$-5 + 6 = -9 + 10$$
$$1 = 1 \quad True$$

47.
$$\tfrac{1}{2}y - 2 = \tfrac{1}{4}y + 3$$
$$\tfrac{1}{2}y - 2 + 2 = \tfrac{1}{4}y + 3 + 2$$
$$\tfrac{1}{2}y = \tfrac{1}{4}y + 5$$
$$\tfrac{1}{2}y - \tfrac{1}{4}y = \tfrac{1}{4}y - \tfrac{1}{4}y + 5$$
$$\tfrac{1}{4}y = 5$$
$$4 \cdot \tfrac{1}{4}y = 5 \cdot 4$$
$$y = 20$$

Check:
$$\tfrac{1}{2}y - 2 = \tfrac{1}{4}y + 3$$
$$\tfrac{1}{2} \cdot 20 - 2 = \tfrac{1}{4} \cdot 20 + 3$$
$$\tfrac{1}{2} \cdot \tfrac{\overset{10}{\cancel{20}}}{1} - 2 = \tfrac{1}{4} \cdot \tfrac{\overset{5}{\cancel{20}}}{1} + 3$$
$$10 - 2 = 5 + 3$$
$$8 = 8 \quad True$$

51.
$$-4(p - 8) + 3(2p + 1) = 7$$
$$-4p + (-4)(-8) + 6p + 3 = 7$$
$$-4p + 32 + 6p + 3 = 7$$
$$-4p + 6p + 32 + 3 = 7$$
$$2p + 35 = 7$$

$$2p + 35 - 35 = 7 - 35$$
$$2p = -28$$
$$\frac{\overset{1}{\cancel{2}}p}{\underset{1}{\cancel{2}}} = \frac{-28}{2}$$
$$p = -14$$

55. $p = 2500$, $r = 11\% = .11$,
$t = 6$ months $= \frac{6}{12}$ year

$I = prt$
$= 2500 \cdot .11 \cdot \frac{6}{12}$
$= 275 \cdot \frac{6}{12}$
$= 275 \cdot \frac{1}{2}$
$= 137.5$

John must pay $137.50 interest.

Section 9.8

3. $\underline{\text{4 added to}} \; \underline{\text{a number}}$
   $\;\;\;4 \quad\quad + \quad\quad\;\; x \quad\;\; = 4 + x$

7. Students often make an error with this problem and write $9 - x$, but that means 9 minus a number. The correct answer is $x - 9$.

11. $\underline{\text{The product of}} \; \underline{\text{a number}} \; \underline{\text{and}} \; \underline{2}$
    $\quad\quad\quad\quad\quad\quad\quad\; x \quad\;\; \cdot \;\; 2 = 2x$

15. A number $\div 2 = x \div 2$ or $\frac{x}{2}$.

19. $\underline{\text{Five times a number}} \; \underline{\text{added to}}$
    $\quad\quad\quad 5x \quad\quad\quad\quad\; +$
    $\underline{\text{four times the number}} = 5x + 4x$
    $\quad\quad\quad 4x$

23. 1st: $x =$ the number

2nd: translate:
$\underline{\text{Twice a number}} \; \underline{\text{is added to}}$
$\quad\quad 2x \quad\quad\quad\quad\quad +$
$\underline{\text{four times the number}}.$
$\quad\quad\quad 4x$
$\underline{\text{The result is 48}}.$
$\quad\quad = 48$

3rd: $2x + 4x = 48$
$(2 + 4)x = 48$
$6x = 48$
$\frac{6x}{6} = \frac{48}{6}$

4th: $\quad\quad\quad\quad x = 8$

5th: Check:
2 times 8 added to 4 times 8 is 48. Correct

27. 1st: $x =$ length of the shorter piece
(then $x + 10$ is length of other piece)

2nd: translate:
$\underline{\text{length of short piece}} \; \underline{\text{and}}$
$\quad\quad\quad x \quad\quad\quad\quad\quad +$
$\underline{\text{length of other piece}} \; \underline{\text{are}}$
$\quad\quad x + 10 \quad\quad\quad\quad =$
$\underline{\text{length of the board}}$
$\quad\quad\quad 78$

3rd: $x + x + 10 = 78$
$2x + 10 = 78$
$2x + 10 - 10 = 78 - 10$
$2x = 68$
$\frac{2x}{2} = \frac{68}{2}$

4th: $\quad\quad\quad x = 34$ cm

5th: Check:
$$34 \text{ cm} + (34 + 10) \text{ cm}$$
$$= 34 \text{ cm} + 44 \text{ cm}$$
$$= 78 \text{ cm} \quad \text{Correct}$$

31. 1st: x = length of rectangle

2nd: translate:
$$P = 2l + 2w$$
$$48 = 2x + 2 \cdot 5$$

3rd: $48 = 2x + 10$
$$48 - 10 = 2x + 10 - 10$$
$$38 = 2x$$
$$\frac{38}{2} = \frac{2x}{2}$$
$$19 = x$$

4th: The length is 19 m.

5th: Check:
$$2 \cdot 19 + 2 \cdot 5$$
$$= 38 + 10$$
$$= 48 \text{ m} \quad \text{Correct}$$

35. 1st: x = number of years

2nd: translate:
$$I = prt$$
$$480 = 800 \cdot .12 \cdot x$$

3rd: $480 = 800 \cdot .12 \cdot x$
$$480 = 96 \cdot x$$
$$\frac{480}{96} = \frac{96 \cdot x}{96}$$
$$5 = x$$

4th: Deposit the money for 5 years.

5th: $I = prt$
$$480 = 800 \cdot .12 \cdot 5$$
$$480 = 96 \cdot 5$$
$$480 = 480 \quad \text{Correct}$$

39. 225 is 75% of what number?

$$\frac{225}{b} = \frac{75}{100}$$

$$\frac{225}{b} = \frac{3}{4} \quad \text{Lowest terms}$$

$$3 \cdot b = 4 \cdot 225$$

$$3 \cdot b = 900$$

$$\frac{\overset{1}{\cancel{3}} \cdot b}{\underset{1}{\cancel{3}}} = \frac{900}{3}$$

$$b = 300$$

225 is 75% of 300.

Chapter 9 Review Exercises

1-4. See answer graphs in the answer section of the textbook.

5. 4 > 2 since 4 is to the right of 2 on a number line.

6. -3 < -1 since -3 is to the left of -1 on a number line.

7. 6 > -9 since 6 is to the right of -9 on a number line.

8. -3 > -5 since -3 is to the right of -5 on a number line.

9. |8| = 8 (8 is 8 units from 0.)

10. |-19| = 19 (-19 is 19 units from 0.)

11. |-7| = 7 (-7 is 7 units from 0.)

12. |0| = 0 (0 is 0 units from 0.)

13. -4 + 6 = 2 (Different signs so subtract absolute values. Use positive sign.)

14. -3 + 12 = 9 (Different signs so subtract absolute values. Use positive sign.)

15. -11 + (-8) = -19 (Same signs so add absolute values. Keep negative sign.)

16. 19 + (-24) = -5 (Different signs so subtract absolute values. Use negative sign.)

7. -7.6 + (-2.1) = -9.7 (Same signs so add absolute values. Keep negative sign.)

18. 8.9 + (-15.7) = -6.8 (Different signs so subtract absolute values. Use negative sign.)

19. $\frac{9}{10} + (-\frac{3}{5})$
    $= \frac{9}{10} + (-\frac{6}{10})$
    $= \frac{3}{10}$

20. $\begin{array}{r} -9 \\ -4 \\ + 6 \\ \hline \end{array} \Big] \begin{array}{r} -13 \\ 6 \\ \hline -7 \end{array}$

21. $\begin{array}{r} -11 \\ -10 \\ + -14 \\ \hline \end{array} \Big] \begin{array}{r} -21 \\ -14 \\ \hline -35 \end{array}$

22. $\begin{array}{r} -12 \\ 8 \\ 9 \\ + -7 \\ \hline \end{array} \Big] \begin{array}{r} -12 \\ 17 \\ -7 \\ \hline \end{array} \Big] \begin{array}{r} 5 \\ -7 \\ \hline -2 \end{array}$

23. $\begin{array}{r} 15 \\ -3 \\ -4 \\ + 12 \\ \hline \end{array} \Big] \begin{array}{r} 15 \\ -7 \\ 12 \\ \hline \end{array} \Big] \begin{array}{r} 8 \\ 12 \\ \hline 20 \end{array}$

24. The additive inverse of 6 is -(6) = -6.

25. The additive inverse of -9 is -(-9) = 9.

26. The additive inverse of $\frac{2}{3}$ is $-\left(\frac{2}{3}\right) = -\frac{2}{3}$.

27. $-(-4) = +4$ or 4

28. $-(-15) = +15$ or 15

29. $-\left(-\frac{5}{8}\right) = +\frac{5}{8}$ or $\frac{5}{8}$

30. $6 - 9 = 6 + (-9)$
    $= -3$

31. $-9 - 2 = -9 + (-2)$
    $= -11$

32. $3 - (-3) = 3 + 3$
    $= 6$

33. $-3 - (-8) = -3 + 8$
    $= 5$

34. $-\frac{3}{4} - \frac{1}{8} = -\frac{3}{4} + \left(-\frac{1}{8}\right)$
    $= -\frac{6}{8} + \left(-\frac{1}{8}\right)$
    $= -\frac{7}{8}$

35. $-4 \cdot 6 = -24$ (Different signs, product is negative.)

36. $5 \cdot (-4) = -20$ (Different signs, product is negative.)

37. $-3 \cdot (-5) = 15$ (Same signs, product is positive.)

38. $-\frac{\overset{1}{2}}{\underset{4}{3}} \cdot \left(-\frac{5}{8}\right) = \frac{5}{12}$ (Same signs, product is positive.)

39. $8.9 \cdot (-4.2) = -37.38$ (Different signs, product is negative.)

40. $\frac{-9}{3} = -3$ (Different signs, quotient is negative.)

41. $\frac{-25}{5} = -5$ (Different signs, quotient is negative.)

42. $\frac{-120}{-6} = 20$ (Same signs, quotient is positive.)

43. $2 \cdot (-5) - 11 = -10 - 11$
    $= -10 + (-11)$
    $= -21$

44. $(-4) \cdot (-8) - 9 = 32 - 9$
    $= 23$

45. $48 \div (-2)^3 - (-5) = 48 \div (-8) + 5$
    $= -6 + 5$
    $= -1$

46. $-36 \div (-3)^2 - (-2) = -36 \div 9 + 2$
    $= -4 + 2$
    $= -2$

47. $5 \cdot 4 - 7 \cdot 6 + 3 \cdot (-4) = 20 - 42 - 12$
    $= -22 - 12$
    $= -34$

48. $2 \cdot 8 - 4 \cdot 9 + 2 \cdot (-6) = 16 - 36 - 12$
$ = -20 - 12$
$ = -32$

49. $-4 \cdot 3^3 - 2 \cdot (5 - 9)$
$= -4 \cdot 27 - 2 \cdot (-4)$
$= -108 + 8$
$= -100$

50. $6 \cdot (-4)^2 - 3 \cdot (7 - 14)$
$= 6 \cdot (16) - 3 \cdot (-7)$
$= 96 + 21$
$= 117$

51. $4^2 \cdot (9 - 16) \div (-5 - 3)$
$= 16 \cdot (-7) \div (-8)$
$= -112 \div (-8)$
$= 14$

52. $(.8)^2 \cdot (.2) - (-1.2)$
$= .64 \cdot (.2) + 1.2$
$= .128 + 1.2$
$= 1.328$

53. $(-.3)^2 \cdot (-.1) - 2.4$
$= .09 \cdot (-.1) - 2.4$
$= -.009 - 2.4$
$= -2.409$

54. $\dfrac{2}{\cancel{9}_3} \cdot \left(-\dfrac{\cancel{6}^2}{7}\right) - \left(-\dfrac{1}{3}\right) = -\dfrac{4}{21} + \dfrac{1}{3}$
$ = -\dfrac{4}{21} + \dfrac{7}{21}$
$ = \dfrac{3}{21}$
$ = \dfrac{1}{7}$

55. $2k + 4m; \ k = 4, \ m = 3$
$= 2(4) + 4(3)$
$= 8 + 12$
$= 20$

56. $2k + 4m; \ k = -6, \ m = 2$
$= 2(-6) + 4(2)$
$= -12 + 8$
$= -4$

57. $2k + 4m; \ k = -8, \ m = -5$
$= 2(-8) + 4(-5)$
$= -16 + (-20)$
$= -36$

58. $2p + q; \ p = -1, \ q = 4$
$= 2(-1) + 4$
$= -2 + 4$
$= 2$

59. $\dfrac{5a - 7y}{2 + m}; \ a = -1, \ y = 4, \ m = -3$
$= \dfrac{5(-1) - 7(4)}{2 + (-3)}$
$= \dfrac{-5 - 28}{-1}$
$= \dfrac{-33}{-1}$
$= 33$

60. $P = a + b + c;$
$a = 9, \ b = 12, \ c = 14$
$= 9 + 12 + 14$
$= 35$

61. $A = \frac{1}{2}bh;\ b = 6,\ h = 9$

$= \frac{1}{2}(6)(9)$

$= \frac{54}{2}$

$= 27$

62. $y + 2 = 11$

$y + 2 - 2 = 11 - 2$

$y + 0 = 9$

$y = 9$

63. $6 = a - 4$

$6 + 4 = a - 4 + 4$

$10 = a + 0$

$10 = a$

64. $z + 5 = 2$

$z + 5 - 5 = 2 - 5$

$z + 0 = -3$

$z = -3$

65. $-8 = -9 + r$

$-8 + 9 = -9 + 9 + r$

$1 = 0 + r$

$1 = r$

66. $x + \frac{1}{6} = \frac{5}{3}$

$x + \frac{1}{6} - \frac{1}{6} = \frac{5}{3} - \frac{1}{6}$

$x + 0 = \frac{10}{6} - \frac{1}{6}$

$x = \frac{9}{6}$

$x = \frac{3}{2} = 1\frac{1}{2}$

67. $12.92 + k = 4.87$

$12.92 - 12.92 + k = 4.87 - 12.92$

$0 + k = -8.05$

$k = -8.05$

68. $8r = 56$

$\frac{8r}{8} = \frac{56}{8}$

$r = 7$

69. $-3p = 24$

$\frac{-3p}{-3} = \frac{24}{-3}$

$p = -8$

70. $\frac{z}{4} = 5$

$4 \cdot \frac{z}{4} = 5 \cdot 4$

$z = 20$

71. $\frac{a}{5} = -11$

$5 \cdot \frac{a}{5} = -11 \cdot 5$

$a = -55$

72. $-\frac{11}{10}y = 22$

$(-\frac{10}{11}) \cdot (-\frac{11}{10}y) = \overset{2}{\cancel{22}} \cdot (-\frac{10}{\underset{1}{\cancel{11}}})$

$y = -20$

73. $-\frac{5}{8}m = \frac{5}{12}$

$(-\frac{8}{5}) \cdot (-\frac{5}{8}m) = (\frac{\overset{1}{\cancel{5}}}{\underset{3}{\cancel{12}}}) \cdot (-\frac{\overset{2}{\cancel{8}}}{\underset{1}{\cancel{5}}})$

$m = -\frac{2}{3}$

74.
$$2y + 6 = 12$$
$$2y + 6 - 6 = 12 - 6$$
$$2y + 0 = 6$$
$$\frac{2y}{2} = \frac{6}{2}$$
$$y = 3$$

75.
$$-3p + 5 = -1$$
$$-3p + 5 - 5 = -1 - 5$$
$$-3p + 0 = -6$$
$$\frac{-3p}{-3} = \frac{-6}{-3}$$
$$p = 2$$

76. $6(r - 5) = 6 \cdot r - 6 \cdot 5$
$ = 6r - 30$

77. $11(p + 7) = 11 \cdot p + 11 \cdot 7$
$ = 11p + 77$

78. $-9(z - 3) = -9 \cdot z - (-9) \cdot 3$
$ = -9z - (-27)$
$ = -9z + 27$

79. $3r + 8r = (3 + 8)r$
$ = 11r$

80. $10z - 15z = (10 - 15)z$
$ = -5z$

81. $3p - 12p = (3 - 12)p$
$ = -9p$

82.
$$4z + 2z = 42$$
$$(4 + 2)z = 42$$
$$6z = 42$$
$$\frac{6z}{6} = \frac{42}{6}$$
$$z = 7$$

Check:
$$4z + 2z = 4z$$
$$4 \cdot 7 + 2 \cdot 7 = 42$$
$$28 + 14 = 42$$
$$42 = 42 \quad \text{True}$$

83.
$$9k - 2k = -35$$
$$(9 - 2)k = -35$$
$$7k = -35$$
$$\frac{7k}{7} = \frac{-35}{7}$$
$$k = -5$$

Check:
$$9k - 2k = -35$$
$$9 \cdot (-5) - 2 \cdot (-5) = -35$$
$$-45 - (-10) = -35$$
$$-45 + 10 = -35$$
$$-35 = -35 \quad \text{True}$$

84.
$$8r - 16r = 24$$
$$(8 - 16)r = 24$$
$$-8r = 24$$
$$\frac{-8r}{-8} = \frac{24}{-8}$$
$$r = -3$$

Check:
$$8r - 16r = 24$$
$$8 \cdot (-3) - 16 \cdot (-3) = 24$$
$$-24 - (-48) = 24$$
$$-24 + 48 = 24$$
$$24 = 24 \quad \text{True}$$

85. <u>18 plus a number</u>
$18 + x = 18 + x$

86. <u>half a number</u>
$\frac{1}{2} \cdot x = \frac{1}{2}x \text{ or } \frac{x}{2}$

87. the sum of four times a number
    $\phantom{\text{the sum of }}$ 4x
    and 6
    6   = 4x + 6

88. three times a number added to 7
    $\phantom{\text{three times}}$ 3x $\phantom{\text{added to}}$ + 7
    = 3x + 7

89. x = the number
    $11x - 8x = -9$
    $(11 - 8)x = -9$
    $3x = -9$
    $x = -3$

90. x = number of additional days rented
    $45 + 35x = 255$
    $45 - 45 + 35x = 255 - 45$
    $0 + 35x = 210$
    $\dfrac{35x}{35} = \dfrac{210}{35}$
    $x = 6$ days
    Total number of days rented is
    $1 + 6 = 7$ days.

91. L = unknown length
    $P = 2L + 2w$
    $124 = 2L + 2 \cdot 25$
    $124 = 2L + 50$
    $124 - 50 = 2L + 50 - 50$
    $74 = 2L$
    $\dfrac{74}{2} = \dfrac{2L}{2}$
    $37 = L$

92. t = unknown time in years
    $I = prt$
    $600 = 1500 \cdot .10 \cdot t \quad 10\% = .10$
    $600 = 150 \cdot t$
    $\dfrac{600}{150} = \dfrac{150t}{150}$
    $4 = t$

93. $-\dfrac{5}{8} \div \left(-\dfrac{3}{16}\right)$ (Same signs, quotient is positive.)
    $= -\dfrac{5}{\overset{1}{\cancel{8}}} \cdot \left(-\dfrac{\overset{2}{\cancel{16}}}{3}\right)$
    $= \dfrac{10}{3} = 3\dfrac{1}{3}$

94. $-7.2 - (-8.6) = -7.2 + 8.6$
    $= 1.4$

95. $3^2 \cdot (7 - 9) \div (-5 - 1)$
    $= 9 \cdot (-2) \div (-6)$
    $= -18 \div (-6)$
    $= 3$

96. $-\dfrac{7}{12} + \dfrac{7}{18} = -\dfrac{21}{36} + \dfrac{14}{36}$
    $= -\dfrac{7}{36}$

97. $|-6| + 2 - (-3) \cdot (-8) + 5$
    $= 6 + 2 - (-3) \cdot (-8) + 5$
    $= 6 + 2 - 24 + 5$
    $= 8 - 24 + 5$
    $= 8 + (-24) + 5$
    $= -16 + 5$
    $= -11$

98. $-\frac{3}{4} \cdot (-\frac{8}{5}) + (-\frac{2}{3})$ (with cancellation: 3/4 cancels to 1, 8/5 cancels to 2)

   $-\frac{6}{5} + (-\frac{2}{3})$

   $-\frac{18}{15} + (-\frac{10}{15})$

   $-\frac{8}{15}$

99. $-\frac{1}{2}y - 3 = -5$

   $-\frac{1}{2}y - 3 + 3 = -5 + 3$

   $-\frac{1}{2}y + 0 = -2$

   $-2 \cdot (-\frac{1}{2}y) = -2 \cdot (-2)$

   $y = 4$

100. $-\frac{2}{3}m + 2 = -2$

   $-\frac{2}{3}m + 2 - 2 = -2 - 2$

   $-\frac{2}{3}m + 0 = -4$

   $-\frac{3}{2} \cdot (-\frac{2}{3}m) = -\frac{3}{2} \cdot (-4)$

   $m = 6$

101. $6z - 3 = 3z + 9$

   $6z - 3z - 3 = 3z - 3z + 9$

   $(6 - 3)z - 3 = 0 + 9$

   $3z - 3 + 3 = 9 + 3$

   $3z + 0 = 12$

   $\frac{3z}{3} = \frac{12}{3}$

   $z = 4$

102. $4 + 2p + 6p = 9 + 4p + 1$

   $4 + 8p = 4p + 10$

   $4 - 4 + 8p = 4p + 10 - 4$

   $8p = 4p + 6$

   $8p - 4p = 4p - 4p + 6$

   $4p = 6$

   $\frac{4p}{4} = \frac{6}{4}$ (reducing to $\frac{1}{1}$ and $\frac{3}{2}$)

   $p = \frac{3}{2}$ or $1\frac{1}{2}$

103. <u>twice</u> <u>a number</u> <u>decreased by</u> <u>8</u>
     2   · x   −   8

   $= 2x - 8$

104. <u>one-third</u> <u>of a number</u> <u>increased by</u> <u>7</u>
     $\frac{1}{3}$ · x   +   7

   $= \frac{1}{3}x + 7$

105. x = unknown width

   (Then $2x + 3$ is the unknown length.)

   $P = 2l + 2w$

   $36 = 2 \cdot (2x + 3) + 2 \cdot x$

   $36 = 4x + 6 + 2x$

   $36 = 6x + 6$

   $36 - 6 = 6x + 6 - 6$

   $30 = 6x$

   $\frac{30}{6} = \frac{6x}{6}$

   $5 = x$

   The width is 5 inches, so the length is $2 \cdot 5 + 3 = 13$ inches.

106. x = price of television

(Then $.05 \cdot x$ is the amount of sales tax, since it represents 5% of x.)

Price of television	plus	amount of sales tax	is	total cost.
x	+	.05x	=	404.25
1.00x	+	.05x	=	404.25
(1.00	+	.05)x	=	404.25
		1.05x	=	404.25

$$\frac{1.05x}{1.05} = \frac{404.25}{1.05}$$

$$x = 385$$

The price of the television is $385.

Chapter 9 Test

1. See graph in the answer section of the textbook.

2. $-9 < -6$

3. $|7| = 7$

4. $-8 + 7 = -1$

5. $-11 + (-2) = -13$

6. $-\frac{3}{8} + (-1\frac{1}{4}) = -\frac{3}{8} + (-\frac{5}{4})$

    $= -\frac{3}{8} + (-\frac{10}{8})$

    $= -\frac{13}{8} = -1\frac{5}{8}$

7. $8 - 15 = 8 + (-15)$

    $= -7$

8. $4 - (-12) = 4 + 12$

    $= 16$

9. $-\frac{1}{2} - (-\frac{3}{4}) = -\frac{1}{2} + \frac{3}{4}$

    $= -\frac{2}{4} + \frac{3}{4}$

    $= \frac{1}{4}$

10. $8(-4) = -32$

11. $-7(-12) = 84$

12. $\frac{-100}{4} = -25$

13. $(-5) \cdot (-3)^2 - (-2) = (-5) \cdot (9) + 2$

    $= -45 + 2$

    $= -43$

14. $(-5) \cdot (3 - 9) - (-4)$

    $= (-5) \cdot (-6) + 4$

    $= 30 + 4$

    $= 34$

15. $8k - 3m; k = -4, m = 2$

    $8(-4) - 3(2) = -32 - 6$

    $= -32 + (-6)$

    $= -38$

16. $8k - 3m; k = 7, m = 9$

    $8(7) - 3(9) = 56 - 27$

    $= 56 + (-27)$

    $= 29$

17. $A = \frac{1}{2}bh$; $b = 20$, $h = 11$

$= \frac{1}{2} \cdot (20) \cdot (11)$

$= 110$

18. $x - 9 = -4$

$x - 9 + 9 = 4 + 9$

$x + 0 = 5$

$x = 5$

19. $-2 + r = 5$

$-2 + 2 + r = 5 + 2$

$0 + r = 7$

$r = 7$

20. $11k = -99$

$\frac{11k}{11} = \frac{-99}{11}$

$k = -9$

21. $-\frac{1}{4}p = 2$

$-4\left(-\frac{1}{4}p\right) = -4(2)$

$p = -8$

22. $8r - 3r = -25$

$(8 - 3)r = -25$

$5r = -25$

$\frac{5r}{5} = \frac{-25}{5}$

$r = -5$

23. $3m - 5 = 7m - 13$

$3m - 7m - 5 = 7m - 7m - 13$

$(3 - 7)m - 5 = 0 - 13$

$-4m - 5 = -13$

$-4m - 5 + 5 = -13 + 5$

$-4m + 0 = -8$

$\frac{-4m}{-4} = \frac{-8}{-4}$

$m = 2$

24. $x =$ the number

$7x + 3 = 17$

$7x + 3 - 3 = 17 - 3$

$7x + 0 = 14$

$\frac{7x}{7} = \frac{14}{7}$

$x = 2$

25. $x =$ smaller piece

$x + 4 =$ larger piece

$x + x + 4 = 118$

$2x + 4 = 118$

$2x + 4 - 4 = 118 - 4$

$2x + 0 = 114$

$\frac{2x}{2} = \frac{114}{2}$

$x = 57$ cm

26. $x =$ rate of interest

$I = prt$

$324 = 1200 \cdot x \cdot 3$

$324 = 1200 \cdot 3 \cdot x$

$324 = 3600 \cdot x$

$\frac{324}{3600} = \frac{3600 \cdot x}{3600}$

$.09 = x$

The rate of interest is .09 or 9%.

Cumulative Review   Chapters 7–9

1. 1 foot = 12 inches

2. 3 feet = 1 yard

3. $2\frac{1}{2}$ pounds $\cdot \frac{16 \text{ ounces}}{1 \text{ pound}} = \frac{5}{6} \cdot \frac{16}{1}$

   $= 40$ ounces

4. 5 yards 2 feet 11 inches
   $+$ 2 yards 1 foot 7 inches
    7 yards 3 feet 18 inches
   $=$ 7 yards 1 yard + 1 foot 6 inches
   $=$ 8 yards 1 foot 6 inches

5. 2   7+4=11
    3̸ weeks 5̸ days 7̸ hours   24+7+31
   $-$ 2 weeks 8 days 9 hours
    3 days 22 hours

6. 4 gallons 3 quarts 1 pint
   $\times 9$
    36 gallons 27 quarts 9 pints
    36 gallons 27 quarts + 4 quarts + 1 pint
   $=$ 36 gallons 31 quarts 1 pint
    36 gallons + 7 gallons 3 quarts + 1 pint
   $=$ 43 gallons 3 quarts 1 pint

7. $$ 6 days 19 hours
   4$\overline{)\text{3 weeks}\text{5 days}\text{28 hours}}$
   3 weeks = + 21 days
    26
    24
    2 days = +48 hours
    76
    76
    0

8. $5978 \text{ mm} \cdot \frac{1 \text{ m}}{1000 \text{ mm}} = \frac{5978 \text{ m}}{1000} = 5.978 \text{ m}$

9. $4317 \text{ g} \cdot \frac{1 \text{ kg}}{1000 \text{ g}} = \frac{4317 \text{ kg}}{1000} = 4.317 \text{ kg}$

10. $2.83 \text{ kl} \cdot \frac{1000 \text{ } \ell}{1 \text{ kl}} = 2.83 \cdot 1000 \text{ } \ell$
    $= 2830 \text{ } \ell$

11. 15 quarts $\cdot$ .946 = 14.2 ℓ

12. 96 feet $\cdot$ .305 = 29.3 m

13. 120 km $\cdot$ .62 = 74.4 miles

14. $C = \frac{5(70 - 32)}{9}$
    $= \frac{5(38)}{9}$
    $= 21°C$

15. $C = \frac{5(420 - 32)}{9}$
    $= \frac{5(388)}{9}$
    $= 216°C$

16. $F = \dfrac{9 \cdot 150}{5} + 32$

    $= \dfrac{1350}{5} + 32$

    $= 270 + 32$

    $= 302°F$

17. $P = 2l + 2w;\ l = 11.8,\ w = 5.7$

    $= 2(11.8) + 2(5.7)$

    $= 23.6 + 11.4$

    $= 35\ m$

    $A = l \cdot w$

    $= 11.8 \cdot 5.7$

    $= 67.3\ m^2$ (rounded)

18. $P = 4s \quad s = 14.8$

    $= 4 \cdot 14.8$

    $= 59.2\ cm$

    $A = s^2$

    $= 14.8^2$

    $= 219.0\ cm^2$ (rounded)

19. $A = b \cdot h \quad b = 7.2,\ h = 15.4$

    $= 7.2 \cdot 15.4$

    $= 110.9\ m^2$ (rounded)

20. $A = \dfrac{1}{2}h(b + B) \quad b = 19.2,\ B = 28.4,\ h = 11.6$

    $= \dfrac{1}{2} \cdot 11.6(19.2 + 28.4)$

    $= \dfrac{1}{2} \cdot 11.6(47.6)$

    $= \dfrac{552.16}{2}$

    $= 276.1\ m^2$ (rounded)

21. $P = a + b + c \quad a = 74.8,\ b = 63.7,\ c = 81.7$

    $= 74.8 + 63.7 + 81.7$

    $= 220.2\ m$

22. $P = a + b + c \quad a = 3.47,\ b = 2.99,\ c = 5.03$

    $= 3.47 + 2.99 + 5.03$

    $= 11.49\ ft$

23. $A = \dfrac{1}{2}bh \quad b = 23.4,\ h = 9.7$

    $= \dfrac{1}{2} \cdot 23.4 \cdot 9.7$

    $= 113.5\ m^2$ (rounded)

24. $A = \dfrac{1}{2}bh \quad b = 892,\ h = 541$

    $= \dfrac{1}{2} \cdot 892 \cdot 541$

    $= 241{,}286\ cm^2$

25. $d = 2r \quad r = 9.48$

    $= 2 \cdot 9.48$

    $= 18.96\ m$

26. $r = \dfrac{d}{2} \quad d = 13.4$

    $= \dfrac{13.4}{2}$

    $= 6.7\ ft$

27. $C = 2\pi r \quad r = 7.6$

    $= 2 \cdot 3.14 \cdot 57.76$

    $= 47.7\ m$ (rounded)

    $A = \pi r^2$

    $= 3.14 \cdot (7.6)^2$

    $= 3.14 \cdot 57.76$

    $= 181.4\ m^2$ (rounded)

28. $C = \pi d \quad d = 143$
   $= 3.14 \cdot 143$
   $= 449.0$ m (rounded)

   $A = \pi r^2 \quad r = \frac{d}{2} = \frac{143}{2} = 71.5$
   $= 3.14 \cdot (71.5)^2$
   $= 3.14 \cdot 5112.25$
   $= 16{,}052.5$ m² (rounded)

29. $A = \frac{1}{2}\pi r^2 \quad r = 32.8$
   $= \frac{1}{2} \cdot 3.14 \cdot (32.8)^2$
   $= \frac{1}{2} \cdot 3.14 \cdot 1075.84$
   $= 1689.1$ m² (rounded)

30. $A = \frac{1}{2}\pi r^2 \quad r = \frac{d}{2} = \frac{7.6}{2} = 3.8$
   $= \frac{1}{2} \cdot 3.14 \cdot (3.8)^2$
   $= \frac{1}{2} \cdot 3.14 \cdot 14.44$
   $= 22.7$ cm² (rounded)

31. A (triangle) $= \frac{1}{2}bh \quad b = 86.2,\ h = 31$
   $= \frac{1}{2} \cdot 86.2 \cdot 31$
   $= 1336.1$ m²

   A (rectangle) $= lw \quad l = 86.2,\ w = 43$
   $= 86.2 \cdot 43$
   $= 3706.6$ m²

   A (total) $= 1336.1 + 3706.6$
   $= 5042.7$ m²

32. A (circle) $= \pi r^2 \quad r = \frac{d}{2} = \frac{12.4}{2} = 6.2$
   $= 3.14 \cdot (6.2)^2$
   $= 3.14 \cdot 38.44$
   $= 120.7$ m² (rounded)

   A (square) $= s^2 \quad s = 8.8$
   $= 8.8^2$
   $= 77.44$ m²

   A (total) $= 120.7 - 77.44$
   $= 43.3$ m² (rounded)

33. A (semicircle) $= \frac{1}{2}\pi r^2 \quad r = 59.1$
   $= \frac{1}{2} \cdot 3.14 \cdot (59.1)^2$
   $= \frac{1}{2} \cdot 3.14 \cdot 3492.81$
   $= 5483.7$ m² (rounded)

   A (square) $= s^2 \quad s = 48.2$
   $= (48.2)^2$
   $= 2323.2$ m² (rounded)

   A (total) $= 5483.7 - 2323.2$
   $= 3160.5$ m²

34. $V = lwh \quad l = 7.2,\ w = 5.8,\ h = 1.4$
   $= 7.2 \cdot 5.8 \cdot 1.4$
   $= 58.5$ m² (rounded)

35. $V = \frac{4}{3}\pi r^3 \quad r = 1.2$
   $= \frac{4}{3} \cdot 3.14 \cdot (1.2)^3$
   $= \frac{4}{3} \cdot 3.14 \cdot 1.728$
   $= 7.2$ ft³ (rounded)

36. $7 > 4$

37. $3 > -4$

38. $-12 > -15$

39. $-8 < -3$

40. $|4| = 4$

41. $|-22| = 22$

42. $|-86| = 86$

43. $|0| = 0$

44. $-2 + 11 = 9$

45. $-7.23 + 15.4 = 8.17$

46. $-\frac{7}{10} + \frac{3}{8} = -\frac{28}{40} + \frac{15}{40}$
    $= -\frac{13}{40}$

47. $\left.\begin{array}{r}-6 \\ -2\end{array}\right] -8$
    $\underline{+\ 5}\quad \underline{\ \ 5}$
    $\qquad\quad -3$

48. $\left.\begin{array}{r}-4 \\ 9\end{array}\right] 5$
    $\underline{+\ -7}\quad \underline{-7}$
    $\qquad\quad -2$

49. $\left.\begin{array}{r}18 \\ -12\end{array}\right] 6 \left.\right] $
    $\qquad 17\quad 17\ \ ] 23$
    $\underline{+\ -3}\qquad\quad \underline{-3}$
    $\qquad\qquad\qquad 20$

50. $\left.\begin{array}{r}-19 \\ 10\end{array}\right] -9$
    $\left.\underline{+\ \begin{array}{r}-11 \\ -5\end{array}}\right]\ \underline{-16}$
    $\qquad\qquad -25$

51. $-(11) = -11$

52. $-(-19) = 19$

53. $-\left(\frac{5}{8}\right) = -\frac{5}{8}$

54. $-(0) = 0$ *Zero is neither positive nor negative.*

55. $8 - 15 = 8 + (-15)$
    $\qquad\quad = -7$

56. $-11 - 6 = -11 + (-6)$
    $\qquad\qquad = -17$

57. $9 - (-6) = 9 + 6$
    $\qquad\qquad = 15$

58. $15 - (-19) = 15 + 19$
    $\qquad\qquad\ = 34$

59. $-\frac{5}{9} - \left(-\frac{1}{3}\right) = -\frac{5}{9} + \frac{1}{3}$
    $\qquad\qquad\quad = -\frac{5}{9} + \frac{3}{9}$
    $\qquad\qquad\quad = -\frac{2}{9}$

60. $-15.7 - 9.8 = -15.7 + (-9.8)$
    $\qquad\qquad\ \ = -25.5$

61. $-9 \cdot 2 = -18$

62. $-15 \cdot 8 = -120$

63. $7 \cdot (-11) = -77$

64. $-5(-9) = 45$

65. $-8(-7) = 56$

66. $(-6.2)(-3) = 18.6$

67. $-\dfrac{\overset{2}{\cancel{8}}}{11} \cdot \dfrac{5}{\underset{1}{\cancel{4}}} = -\dfrac{10}{11}$

68. $\left(-\dfrac{\overset{1}{\cancel{3}}}{\underset{2}{\cancel{10}}}\right)\left(-\dfrac{\overset{1}{\cancel{5}}}{\underset{3}{\cancel{9}}}\right) = \dfrac{1}{6}$

69. $\dfrac{-28}{7} = -4$

70. $\dfrac{-15}{-3} = 5$

71. $\dfrac{-540}{27} = -20$

72. $\dfrac{\frac{7}{15}}{\frac{14}{3}} = \left(-\dfrac{\overset{1}{\cancel{7}}}{\underset{5}{\cancel{15}}}\right)\left(-\dfrac{\overset{1}{\cancel{3}}}{\underset{2}{\cancel{14}}}\right) = \dfrac{1}{10}$

73. $-2 \cdot (-8) - (-3) = 16 + 3$
$ = 19$

74. $(-7)^2 \cdot (-3) = 49 \cdot (-3)$
$ = -147$

75. $(-.2)^2 \cdot (8 - 15) = .04 \cdot (-7)$
$ = -.28$

76. $7x - 5y \quad x = 2, \; y = 1$
$ = 7(2) - 5(1)$
$ = 14 - 5$
$ = 9$

77. $7x - 5y \quad x = -4, \; y = -3$
$ = 7(-4) - 5(-3)$
$ = -28 + 15$
$ = -13$

78. $8a + b \quad a = 4, \; b = -7$
$ = 8(4) + (-7)$
$ = 32 + (-7)$
$ = 25$

79. $5r - 3s + 4t \quad r = 1, \; s = -2, \; t = -1$
$ = 5(1) - 3(-2) + 4(-1)$
$ = 5 + 6 + (-4)$
$ = 7$

80. $P = 2l + 2w \quad l = 17.4, \; w = 13.9$
$ = 2(17.4) + 2(13.9)$
$ = 34.8 + 27.8$
$ = 62.6$

81. $V = \dfrac{1}{3}bh \quad b = 540, \; h = 32$
$ = \dfrac{1}{3} \cdot 540 \cdot 32$
$ = 5760$

82. $ m + 7 = 15$
$m + 7 - 7 = 15 - 7$
$ m + 0 = 8$
$ m = 8$
Check:
$ 8 + 7 = 15$
$ 15 = 15$

83. $12 = r - 3$
     $12 + 3 = r - 3 + 3$
     $15 = r + 0$
     $15 = r$
   Check:
     $12 = 15 - 3$
     $12 = 12$

84. $a + 6 = 2$
     $a + 6 - 6 = 2 - 6$
     $a + 0 = -4$
     $a = -4$
   Check:
     $-4 + 6 = 2$
     $2 = 2$

85. $a + \frac{2}{3} = \frac{5}{12}$
     $a + \frac{2}{3} - \frac{2}{3} = \frac{5}{12} - \frac{2}{3}$
     $a + 0 = \frac{5}{12} - \frac{8}{12}$
     $a = -\frac{3}{12}$
     $a = -\frac{1}{4}$
   Check:
     $-\frac{1}{4} + \frac{2}{3} = \frac{5}{12}$
     $-\frac{3}{12} + \frac{8}{12} = \frac{5}{12}$
     $\frac{5}{12} = \frac{5}{12}$

86. $12m = 72$
     $\frac{12m}{12} = \frac{72}{12}$
     $m = 6$
   Check:
     $12 \cdot 6 = 72$
     $72 = 72$

87. $\frac{k}{2} = 13$
     $\frac{2}{1} \cdot \frac{k}{2} = 13 \cdot \frac{2}{1}$
     $k = 26$
   Check:
     $\frac{26}{2} = 13$
     $13 = 13$

88. $\frac{3}{5}p = -12$
     $\frac{5}{3} \cdot \frac{3}{5}p = -12 \cdot \frac{5}{3}$
     $p = -20$
   Check:
     $\frac{3}{5}(-20) = -12$
     $-12 = -12$

89. $-\frac{5}{9}z = 10$
     $\left(-\frac{9}{5}\right)\left(-\frac{5}{9}z\right) = 10\left(-\frac{9}{5}\right)$
     $z = -18$
   Check:
     $-\frac{5}{9} \cdot (-18) = 10$
     $10 = 10$

90. $-\frac{2}{3}r = -6$

$\left(-\frac{3}{2}\right)\left(-\frac{2}{3}r\right) = -6\left(-\frac{3}{2}\right)$

$r = 9$

Check:

$-\frac{2}{3}(9) = -6$

$-6 = -6$

91. $-\frac{7}{8}k = \frac{3}{5}$

$\left(-\frac{8}{7}\right)\left(-\frac{7}{8}\right) = \frac{3}{5}\left(-\frac{8}{7}\right)$

$k = -\frac{24}{35}$

Check:

$-\frac{7}{8}\left(-\frac{24}{35}\right) = \frac{3}{5}$

$\frac{3}{5} = \frac{3}{5}$

92. $3r + 2 = 8$

$3r + 2 - 2 = 8 - 2$

$3r + 0 = 6$

$\frac{3r}{3} = \frac{6}{3}$

$r = 2$

Check:

$3(2) + 2 = 8$

$6 + 2 = 8$

$8 = 8$

93. $-\frac{1}{2}(a) + 3 = 5$

$-\frac{1}{2}(a) + 3 - 3 = 5 - 3$

$-\frac{1}{2}a + 0 = 2$

$\left(-\frac{2}{1}\right)\left(-\frac{1}{2}a\right) = 2\left(-\frac{2}{1}\right)$

$a = -4$

Check:

$-\frac{1}{2}(-4) + 3 = 5$

$2 + 3 = 5$

$5 = 5$

94. $2p + 5p = -49$

$(2 + 5)p = -49$

$7p = -49$

$\frac{7p}{7} = \frac{-49}{7}$

$p = -7$

Check:

$2(-7) + 5(-7) = -49$

$-14 - 35 = -49$

$-49 = -49$

95. $8a + 7 = 3a + 2$

$8a + 7 - 7 = 3a + 2 - 7$

$8a + 0 = 3a - 5$

$8a - 3a = 3a - 3a - 5$

$(8 - 3)a = 0 - 5$

$5a = -5$

$\frac{5a}{5} = \frac{-5}{a}$

$a = -1$

Check:

$8(-1) + 7 = 3(-1) + 2$

$-8 + 7 = -3 + 2$

$-1 = -1$

96. $4(a + 6) = 4a + 4 \cdot 6$
    $= 4a + 24$

97. $-4(m - 3) = -4 \cdot m + (-4)(-3)$
    $= -4m + 12$

98. $-6(r - 7) = -6 \cdot r + (-6)(-7)$
    $= -6r + 42$

99. <u>seven times a number</u>, <u>added to -12</u>,
            $7 \cdot x$ $+ (-12)$
    <u>result is 23</u>
    $= 23$
    $7x + (-12) = 23$
    $7x - 12 + 12 = 23 + 12$
    $7x + 0 = 35$
    $\dfrac{7x}{7} = \dfrac{35}{7}$
    $x = 5$

100. 1st piece $= x$
     2nd piece $= x + 23$
     $x + (x + 23) = 167$
     $2x + 23 = 167$
     $2x + 23 - 23 = 167 - 23$
     $2x + 0 = 144$
     $\dfrac{2x}{2} = \dfrac{144}{2}$
     $x = 72$ cm

CHAPTER 10 STATISTICS
Section 10.1

3. $\dfrac{\text{cost of appliances}}{\text{total cost}} = \dfrac{3800}{10,400} = \dfrac{19}{52}$

7. History, at 700, has the least students.

11. $\dfrac{\text{History}}{\text{Social science}} = \dfrac{700}{2500} = \dfrac{7}{25}$

15. Entertainment is 10% of $580,000.
    $= .10 \cdot 580,000$
    $= \$58,000$

19. Supplies are 20% of $19,600.
    $= .20 \cdot 19,600$
    $= \$3920$

23. National dues are 10% of $19,600.
    $= .10 \cdot 19,600$
    $= \$1960$

27. Books: 10% of 360°
    $= .10 \cdot 360$
    $= 36°$

31. Food   = 90°
    Rent   = 72°
    Clothing   = 54°
    Books   = 36°
    Entertainment = 54°
    Savings   = 18°
    Other   = + 36°
                   ─────
                    360°

    See the circle graph in the answer section of the textbook.

35. $\dfrac{1}{4}$ ─── $\dfrac{3}{8}$

    $\dfrac{2}{8} < \dfrac{3}{8}$

    $\dfrac{1}{4} < \dfrac{3}{8}$

39. $38.25\% \leq 38.29\%$
    Both numbers are percents, so all that is necessary is to observe that 38.25 is smaller than 38.29.

Section 10.2

3. July 3 attendance was 2000.

7. May has the greatest number employed. The total was 10 hundred or 1000.

11. The increase from February 1990 to April 1990 was from 500 to 800 employees.
    Employee increase = 800 − 500 = 300

15. The greatest difference occurred in 1987 with unleaded at 400 and supreme unleaded at 200.

19. The greatest number of burglaries, 600, occurred in April.

23. The burglaries went from 400 in March to 600 in April, an increase of 200.

27. Annual sales for Store A (solid line), in 1989, were 1500 thousand or $1,500,000.

31. Total sales in 1991 were $40,000.

35. The profit in 1990 was $5000.

39. The county receives 39% of the amount collected.
    = 39% of 8400
    = .39 · 8400
    = $3276

Section 10.3

3. Number of members under 30 is:

1	(15 – 20)
2	(20 – 25)
+ 4	(25 – 30)
7 members.	

7. The greatest number of employees are in the 20–25 thousand dollar group. ($20,000 to $25,000)

11. Employees earning $25,000 or less are:

3	(5 – 10)
6	(10 – 15)
7	(15 – 20)
+ 12	(20 – 25)
28 employees.	

15.
Number of sets	Tally
140–149	|||||	

    Frequency
    6

19.
Temperature	Tally
70°–74°	||

    Frequency
    2

23. 90°–94°, |||||, 5

27. 110°–114°, ||, 2

31.
Score	Tally	Frequency
50–59	||||| ||||| |||	13

35.
90–99	||||| |||	8

39. $(5 \cdot 2) + (7 \cdot 3) \div 5$
    = 10 + 21 ÷ 5   *Parentheses first*
    = 10 + 4.2   *Division before addition*
    = 14.2   *Add last*

Section 10.4

3. mean = $\dfrac{\text{sum of all values}}{\text{number of values}}$

    = $\dfrac{40 + 51 + 59 + 62 + 68 + 73 + 49 + 80}{8}$

    = $\dfrac{482}{8}$

    = 60.3 (rounded)

7. Note that there were ten purchases and

    $ 9.40
    11.30
    10.50
    7.40
    9.10
    8.40
    9.70
    5.20
    1.10
    + 4.70
    $76.80 was the sum of all the purchases.

$$\text{mean} = \frac{\text{sum of all purchases}}{\text{number of purchases}}$$

$$= \frac{76.80}{10}$$

$$= \$7.68$$

11.
Value	Frequency	Product
12	4	48
13	2	26
15	5	75
19	3	57
22	1	22
23	5	115
Totals	20	343

Weighted mean = $\frac{343}{20}$ = 17.2

15. 100, 114, 125, 135, 150, 172

The median is the mean of the middle two numbers.

$$\frac{125 + 135}{2} = 130$$

The median is 130.

19. 74, 68, 68, 68, 75, 75, 74, 74, 70

68 and 74 are the modes because they occur most often (three times each).

23. 21, 32, 38, 46, 49, 53, 58, 72, 97

When the numbers are listed in increasing order, 49 is the single number in the middle of the list, so it is the median.

27. x = amount of interest

I = prt

x = 8500 · .12 · 1

x = 1020 · 1

x = 1020

She will earn $1020 interest.

Chapter 10 Review Exercises

1. Largest expense was motels at $280.

2. Total cost was

 280 + 200 + 80 + 140 + 150 = $850.

$$\frac{\text{food}}{\text{total cost}} = \frac{200}{850} = \frac{4}{17}$$

3. $\frac{\text{gasoline}}{\text{total cost}} = \frac{150}{850} = \frac{3}{17}$

4. $\frac{\text{sightseeing}}{\text{total cost}} = \frac{140}{850} = \frac{14}{85}$

5. $\frac{\text{other}}{\text{gasoline}} = \frac{80}{150} = \frac{8}{15}$

6. $\frac{\text{food}}{\text{motels}} = \frac{200}{280} = \frac{5}{7}$

7. Uniforms are 25% of $88,500.

 = .25 · 88,500

 = $22,125

8. Paint and chemicals are 5% of $88,500.

 = .05 · 88,500

 = $4425

9. Athletic equipment is 31% of $88,500.
   = .31 · 88,500
   = $27,435

10. First aid supplies are 19% of $88,500.
    = .10 · 88,500
    = $8850

11. Miscellaneous is 11% of $88,500.
    = .11 · 88,500
    = $9735

12. Grounds care equipment is 18% of $88,500.
    = .18 · 88,500
    = $15,930

13. In 1991 (shaded bar) the greatest amount of water was in March.

14. In 1990 the least amount of water was in June.

15. In March of 1991 there were 8 million acre-feet of water.

16. In May of 1990 there were 4 million acre-feet of water.

17. From March, 1990 to June, 1990 the water went from 7 to 2 million acre-feet, a decrease of 5 million acre-feet of water.

18. From April, 1991 to June, 1991 the water went from 7 to 5 million acre-feet, a decrease of 2 million acre-feet of water.

19. In 1991 district A (solid line) purchased $25,000 worth of books.

20. In 1990 district A (solid line) purchased $20,000 worth of books.

21. In 1989 district A (solid line) purchased $35,000 worth of books.

22. In 1991 district B (dotted line) purchased $40,000 worth of books.

23. In 1990 district B (dotted line) purchased $30,000 worth of books.

24. In 1989 district B (dotted line) purchased $20,000 worth of books.

25. mean
    $$= \frac{6+4+5+8+3+14+18+29+7+1}{10}$$
    $$= \frac{95}{10} = 9.5$$

26. mean
    $$= \frac{31+9+8+22+46+51+48+42+53+42}{10}$$
    $$= \frac{352}{10} = 35.2$$

190 Chapter 10 Statistics

27.

Value	Frequency	Product
42	3	126
47	7	329
53	2	106
55	3	165
59	5	295
Totals	20	1021

weighted mean = $\frac{1021}{20}$ = 51.1

28.

Value	Frequency	Product
243	1	243
247	3	741
251	5	1255
255	7	1785
263	4	1052
271	2	542
279	2	558
Totals	24	6176

weighted mean = $\frac{6176}{24}$ = 257.3

29. Place the nine numbers in numerical order.

27, 29, 29, 32, 34, 42, 48, 51, 74

Median is the 5th number or 34.

30. Place the eight numbers in numerical order.

525, 542, 551, 559, 565, 576, 578, 590

The median is the mean of the two middle numbers.

median = $\frac{559 + 565}{2}$ = 562

31. The mode is 64. It occurs most often.

32. The mode is 29. It occurs most often.

33. Spices: 10% of 360°
= .10 · 360
= 36°

34. Equipment: $\frac{1960}{5600}$ = .35 = 35%

35. Books: $\frac{1120}{5600}$ = .2 = 20%

36. Supplies: 10% of 360
= .10 · 360
= 36°

37. Small appliances: $\frac{1400}{5600}$ = .25 = 25%

38. See the circle graph in the answer section of the textbook.

   Spices   = 36°
   Equipment   = 126°
   Books   = 72°
   Supplies   = 36°
   Small appliances = 90°

39. mean = $\frac{12 + 18 + 13 + 37 + 45}{5}$

    = $\frac{125}{5}$ = 25

40. mean
$$= \frac{122+135+146+159+128+147+168+139+158}{9}$$
$$= \frac{1302}{9} = 144.7 \text{ (rounded)}$$

41. The mode is 97. It occurs most often.

42. There are two modes, 62 and 83.

43. Arrange the numbers in numerical order.
    1.0, 2.9, 3.2, 4.7, 5.3, 7.1, 8.2, 9.4

    The median is the mean of the 4th and 5th numbers.
    $$\text{mean} = \frac{4.7 + 5.3}{2} = 5$$

44. Arrange the numbers in numerical order.
    1, 2, 3, 7, 9, 14, 15, 18, 21, 28, 46, 59
    $$\text{median} = \frac{14 + 15}{2} = 14.5$$

	Score	Tally	Frequency
45.	30–39	||||	4
46.	40–49	|	1
47.	50–59	⧸⧹⧹⧹ |	6
48.	60–69	||	2
49.	70–79	⧸⧹⧹⧹ ⧸⧹⧹⧹ |||	13
50.	80–89	⧸⧹⧹⧹ ||	7
51.	90–99	⧸⧹⧹⧹ ||	7

52. See the frequency polygon in the answer section of the textbook.

53.
Value	Frequency	Product
12	1	12
14	6	84
17	2	34
23	7	161
25	4	100
Totals	20	391

weighted mean $= \frac{391}{20} = 19.6$ (rounded)

54.
Value	Frequency	Product
104	6	624
112	14	1568
115	21	2415
119	13	1547
123	22	2706
127	6	762
132	9	1188
Totals	91	10,810

weighted mean $= \frac{10,810}{91}$
$= 118.8$ (rounded)

Chapter 10 Test

1. Salaries:
   22% of $2,800,000
   $= .22 \cdot 2,800,000$
   $= \$616,000$

2. Materials:
   18% of $2,800,000
   $= .18 \cdot 2,800,000$
   $= \$504,000$

192   Chapter 10   Statistics

3. Equipment:
   30% of $2,800,000
   = .30 · 2,800,000
   = $840,000

4. Repairs:
   16% of $2,800,000
   = .16 · 2,800,000
   = $448,000

5. Miscellaneous:
   2% of $2,800,000
   = .02 · 2,800,000
   = $56,000

6. Supplies:
   12% of $2,800,000
   = .12 · 2,800,000
   = $336,000

7. Newsprint:
   20% of 360°
   = .20 · 360
   = 72°

8. Ink: $\frac{36°}{360°} = .1 = 10\%$

9. Wire Service:
   30% of 360° = .30 · 360
   = 108°

10. Salaries:
    30% of 360° = .30 · 360
    = 108°

11. Other:
    10% of 360° = .10 · 360
    = 36°

12. See the circle graph in the answer section of the textbook.

	Profit	Number of weeks	Frequency
13.	$120–$129	|||	3
14.	$130–$139	||	2
15.	$140–$149	||||	4
16.	$150–$159	|||	3
17.	$160–$169	|||	3
18.	$170–$179	|||||	5

19. See the histogram in the answer section of your textbook.

20. mean
    $= \frac{42 + 51 + 58 + 59 + 63 + 65 + 69 + 74 + 81 + 88}{10}$
    $= \frac{650}{10} = 65$

21. mean
    $= \frac{12 + 18 + 14 + 17 + 19 + 22 + 23 + 25}{8}$
    $= \frac{150}{8} = 18.8$ (rounded)

22. mean
    $= \frac{458+432+496+491+500+508+512+396+492+504}{10}$
    $= \frac{4789}{10} = 478.9$

23.

Value	Frequency	Product
6	7	42
10	3	30
11	4	44
14	2	28
19	3	57
24	1	24
Totals	20	225

weighted mean = $\frac{225}{20}$ = 11.3

24.

Value	Frequency	Product
150	15	2250
160	17	2720
170	21	3570
180	28	5040
190	19	3610
200	7	1400
Totals	107	18,590

weighted mean = $\frac{18,590}{107}$

= 173.7 (rounded)

25. 15, 18, 19, 27, 29, 31, 42

median = 27

26. 38, 38, 39, 41, 42, 45, 47, 51

median = $\frac{41 + 42}{2}$ = $\frac{83}{2}$ = 41.5

27. 4.2, 5.3, 7.1, 7.6, 8.3, 9.0, 9.3, 10.4, 21.8

median = 8.3

28. The number which occurs most often is 47, so it is the mode.

29. There is no number occuring more than once so there is no mode.

30. Both 103 and 104 are modes.

CHAPTER 11 CONSUMER MATHEMATICS
Section 11.1

3. $4000 at 5% for 9 years
   1st: From table:
       1.5513   *Nine periods at 5%*
   2nd: Multiply:
       4000 · 1.5513   *Multiply by the principal*
       = 6205.2
   Answer: $6205.20

7. $1000 at 8% compounded semiannually for 9 years
   1st: 9 · 2 = 18 periods
         8% ÷ 2 = 4% per period
   2nd: From table: 2.0258
   3rd: Multiply:
       1000 · 2.0258 = 2025.8
   Answer: $2025.80

11. $2800 at 8% compounded quarterly for 5 years
    1st: 5 · 4 = 20 periods
          8% ÷ 4 = 2% per period
    2nd: From table: 1.4859
    3rd: Multiply:
        2800 · 1.4859 = 4160.52
    Answer: $4160.52

15. $1372.80 − $1000 = $372.80 compound interest

19. $1480 at 8% compounded quarterly for 4 years
    1st: 4 · 4 = 16 periods
          8% ÷ 4 = 2% per period
    2nd: From table: 1.3728
    3rd: Multiply:
        1480 · 1.3728 = 2031.744
    Answer: $2031.74 compound amount
    2031.74 − 1480 = $551.74 compound interest

23. (a) $1000 at 12% compounded annually for 3 years
    1st: From table: 1.4049
    2nd: Multiply:
        1000 · 1.4049 = 1404.9
    Answer: $1404.90

    (b) $1000 at 12% compounded semiannually for 3 years
    1st: 3 · 2 = 6 periods
          12% ÷ 2 = 6% per period
    2nd: From table: 1.4185
    3rd: Multiply:
        1000 · 1.4185 = 1418.5
    Answer: $1418.50

    (c) $1000 at 12% compounded quarterly for 3 years
    1st: 3 · 4 = 12 periods
          12% ÷ 4 = 3% per period
    2nd: From table: 1.4258
    3rd: Multiply:
        1000 · 1.4258 = 1425.8
    Answer: $1425.80

27. Quarterly compounding gives the highest return.

31. (a) Bank 1: $10,000 = Principal
    8% = annual rate
    6 = number of periods
    1.5869 = factor from table
    Compound amount
    $= 10{,}000 \times 1.5869$
    $= \$15{,}869$

    Bank 2: $10,000 = Principal
    8%/4 = 2% = quarterly rate
    6 × 4 = 24 = number of periods
    1.6084 = factor from table
    Compound amount
    $= 10{,}000 \times 1.6084$
    $= \$16{,}084$

    (b) Bank 2 pays 16,084 − 15,869 = $215 more interest.

35. $6.3 \div 4.2 \cdot 3.1$
    $= 1.5 \cdot 3.1$ *Divide first*
    $= 4.65$ *Multiply*

39. $4.34 - 2.6 \cdot 5.2 \div 2.6$
    $= 4.34 - 13.52 \div 2.6$ *Multiply first*
    $= 4.34 - 5.2$ *Divide next*
    $= 4.34 + (-5.2)$ *Subtract last*
    $= -.86$

Section 11.2

3. 1st: Amount paid = 243.75 · 24
            = $5850
   2nd: Interest charge
            = 5850 − 5000 = $850
   3rd: $\dfrac{850}{5000} \cdot 100 = \$17$
   4th: From table:
         (24 months and $16.94) = $15\frac{1}{2}$%

7. 1st: Amount paid = 162.61 · 30
            = $4878.30
   2nd: Interest charge
            = 4878.30 − 3950 = $928.30
   3rd: $\dfrac{928.30}{3950} \cdot 100 = \$23.50$
   4th: From table:
         (30 months and $23.45) = 17%

11. $\dfrac{\$92}{\$1000} \cdot 100 = \$9.20$, number of payments = 12
    From table:
    (12 months and $9.16) = $16\frac{1}{2}$%

15. Retail credit: 18%

19. 1st: Note that 25% of $400 is $100. *Down payment*
        27.35 · 12 = 328.20 *Total payments*
        100 + 328.20 = $428.20 *Total amount paid*
    2nd: Interest charge
            = 428.20 − 400 = $28.20

3rd: $\frac{28.20}{300} \cdot 100 = \$9.40$ Finance charge per $100

4th: From table:
(12 months and $9.45) = 17%
(Since there is a down payment of $100, interest is paid only on the $300 borrowed, and that is why the denominator in the 3rd step is 300 instead of 400.)

23. $728 \cdot 100 = 72{,}800$
(When multiplying a whole number by 100, just attach two zeros to the right of the original number.)

27. $\$27.90 \cdot 1000 = \$27{,}900$
(When multiplying a number with a decimal point by 1000, just move the decimal point three places to the right; note that doing so in this problem required thinking of 27.90 as 27.900.)

Section 11.3

3. $30,000 at 12% for 25 years
   1st: From table: 10.53
   2nd: Multiply: $30 \cdot 10.53 = 315.9$
   Answer: $314.90

7. From the table:
   12% interest for 25 years is 10.53 so $134.5 \cdot 10.53 = \$1416.29$ is the amount of principal and interest to be paid each month.

$\frac{\$660}{12} = \$55$ are the monthly taxes, and

$\frac{\$312}{12} = \$26$ is the monthly insurance.

The total monthly payment is

$\$1416.29 + \$55 + \$26 = \$1497.29$.

11. Monthly income = $1720
    Monthly payments = $120
    $1720 - $120 = $1600
    $\frac{1600}{4} = \$400$ maximum house payment

15. Monthly income = $3200
    Monthly payments = $400
    $3200 - $400 = $2800
    $\frac{2800}{4} = \$700$ maximum house payment

19. 1st: The total for principal and interest = amount per payment times number of payments, or $856.80 \cdot 360 = \$308{,}448.00$.

    2nd: Total interest = total amount paid minus amount financed, or $308,448 - $90,000 = $218.448.

23. $11.63   2 decimal places
    ×   15   + 0 decimal places
       58 15   2 decimal places
       116 3   in product
       $174.45

27. $42.10   2 decimal places
    ×   30.5   + 1 decimal places
       21 050   3 decimal places
       00 00   in product
       1263 0
       $1284.050 or $1284.05

Section 11.4

3. $40,000 10-year term, male, age 25

   1st: From table:

       5.77   *Use age and type of policy*

   2nd: Multiply:

       40 · 5.77 = 230.8

       Multiply by the number of thousands

   Annual premium = $230.80

7. $26,500 whole life, female, age 24

   Use row 21 of table since

   24 − 3 = 21.   *Subtract 3 years for female*

   1st: From table: 12.33

       Use age and type of policy

   2nd: Multiply:

       26.5 · 12.33 = 326.745

       Multiply by the number of thousands

   Annual premium = $326.75 (rounded)

11. $40,000 endowment, male, age 30

    1st: From table:

        34.45   *Use age and type of policy*

    2nd: Multiply:

        40 · 34.45 = 1378 *Multiply by the number of thousands*

    Annual premium: $1378

15. $32,000 whole life, male, age 45

    1st: From table:

        23.95   *Use age and type of policy*

    2nd: Multiply:

        32 · 23.95 = 766.4 *Multiply by the number of thousands*

    Annual premium: $766.40

Chapter 11 Review Exercises

1. $500 at 6% for 2 years

   From table:

   1.1236   *Two periods at 6%*

   $500 · 1.1236 = $561.80

       Multiply by the principal

2. $1500 at 8% for 10 years

   From table:

   2.1589   *10 periods at 8%*

   $1500 · 2.1589 = $3238.35

       Multiply by the principal

3. $2150 at $5\frac{1}{2}$% for 20 years

   From table:

   2.9178 *20 periods at $5\frac{1}{2}$%*

   $2150 · 2.9178 = $6273.27

4. $6380.50 at 6% for 15 years

   From table:

   2.3966 *Fifteen periods at 6%*

   $6380.50 · 2.3966 = $15,291.51

5. $400 at 8% semiannually for 2 years

   From table:

   8% ÷ 2 = 4% per period

   2 · 2 = 4 periods

198   Chapter 11   Consumer Mathematics

1.1699

$400 · 1.1699 = $467.96   final amount

$467.96 − $400 = $67.96 compound interest

6. $850 at 12% quarterly for 3 years

From table:

12% ÷ 3 = 4% per period

3 · 4 = 12 periods

1.4258

$850 · 1.4258 = $1211.93   final amount

$1211.93 − $850 = $361.93 compound interest

7. $15,000 at 8% quarterly for 6 years

From table:

8% ÷ 4 = 2% per period

6 · 4 = 24 periods

1.6084

$15,000 · 1.6084 = $24,126

8. Amount paid $54 · 12 = $648

Interest charge = $648 − $600 = $48

Charge per $100 = $\frac{48}{600}$ · 100 = $8

From table:

(12 payments and $8.03) = $14\frac{1}{2}$%

9. Amount paid = $85 · 48 = $4080

Interest charge = $4080 − $3000
= $1080

Charge per $100 = $\frac{1080}{3000}$ · 100 = $36

From table:

(48 payments and $35.03) = 16%

10. Amount paid = $43.10 · 12 + $75
= $517.20 + $75
= $592.20

Interest charge:

$592.20 − $550 = $42.20

Charge per $100 = $\frac{42.20}{475}$ · 100

= $8.88

From table:

(12 payments and $8.88) = 16%

11. The table value at 13% interest for 20 years is 11.72 so

135 · 11.72 = $1582.20.

12. The table value at $11\frac{1}{2}$% interest for 25 years is 10.17 so

82.75 · 10.17 = $841.57.

13. From table:

(10% interest for 20 years) = $9.66 per $1000

Monthly payment:

9.66 · 90   = $869.40

Taxes = 108 ÷ 12   =   9.00

Insurance = 204 ÷ 12 =   17.00

Total   = $895.40

14. From table:

(12% interest for 20 years)
= $11.01 per $1000

Monthly payment:

11.01 · 47.5   = $522.98

Taxes = 252 ÷ 12   =   21.00

Insurance = 180 ÷ 12 =   15.00

Total   = $558.98

15. From table:

   ($11\frac{1}{2}$% interest for 25 years)

   = $10.17 per $1000

   Monthly payment = 10.17 · 75

   = $762.75

16. Total payments = 25 · 12 = 300

17. Total amount = 762.75 · 300

   = $228,825

18. Total interest = $228,825 − $75,000

   = $153,825

19. Total interest paid is more.

20. From table:

   (whole life, age 22)   *Subtract 3 years from age of female*

   = $17.98 per $1000

   Annual premium = 17.98 · 30 = $539.40

21. From table:

   (endowment, age 50)   *Subtract 3 years from age of female*

   = $41.04 per $1000

   Annual premium = 41.04 · 40

   = $1641.60

22. From table:

   (10 year term, age 40)

   = $9.13 per $1000

   Annual premium = 9.13 · 50 = $456.50

23. From table:

   (whole life, age 25)   *Subtract 3 years from age of female*

   = $13.36 per $1000

   Annual premium = $13.36 · 25 = $334

24. From table:

   (endowment, age 35)

   = $35.22 per $1000

   Annual premium = 35.22 · 40 = $1408.80

25. From table:

   (whole life, age 35)   *Subtract 3 years from age of female*

   = $17.44 per $1000

   Annual premium = 17.44 · 40 = $697.60

26. $8225 at 8% quarterly for 6 years

   From table:

   (8% ÷ 4 = 2%, 4 · 6 = 24 periods)

   1.6084

   (a) The compound amount is

   $8225 · 1.6084 = $13,229.09.

   (b) The compound interest is

   $13,229.09 − $8225 = $5004.09.

27. (a) Fair Oaks Building and Thrift:

   $5000 at 8% quarterly for 5 years

   From table:

   (2% interest for 20 years)

   = 1.4859

   $5000 · 1.4859 = $7429.50

   Carmichael Federal Loan:

   $5000 at 8% semiannually for 5 years

From table:
(4% interest for 10 years)
= 1.4802
$5000 · 1.4802 = $7401

(b) She will have $28.50 more in Fair Oaks Building and Thrift.

28. 1st: Amount paid = 325.22 · 30
= $9756.60

2nd: Interest charge
= 9756.60 − 7900
= $1856.60

3rd: $\frac{1856.60}{7900} \cdot 100 = \$23.50$

4th: From table:
(30 months and $23.45) = 17%

29. From table:
($10\frac{1}{2}$% interest for 30 years)
= $9.15 per $1000
Monthly payment:
9.15 · 84.5   = $773.18
Taxes = 1020 ÷ 12 =   85.00
Insurance = 324 ÷ 12 = 27.00
Total   = $885.18

30. From the table, 10% interest for 30 years is 8.78 so
95.5 · 8.78 = $838.49.

31. Amount paid = $177.32 · 18 = $3191.76
(a) Interest charge
= $3191.76 − $2800
= $391.76

(b) Charge per $100 = $\frac{391.76}{2800} \cdot 100$
= $13.99
From table:
(18 months and $13.99) = 17%

32. From table:
(10-year term, age 45)
= $10.82 per $1000
Annual premium = $10.82 · 70
= $757.40.

Chapter 11 Test

1. Table value:
1.8983 Use 11 periods and 6%
Compound amount = 1.8983 · 1000
= $1898.30

2. 7 · 2 = 14 periods
8% ÷ 2 = 4% per period
Table value: 1.7317
Compound amount = 1.7317 · 8500
= $14,719.45

3. 4 · 4 = 16 periods
12% ÷ 4 = 3% per period
Table value: 1.6047
Compound amount = 1.6047 · 15,000
= $24,070.50

4. Amount paid = 18 · $38 = $684
Interest charge = $684 − $600 = $84
Charge per $100 = $\frac{84}{600} \cdot 100 = \$14$
From table:
(18 payments and $13.99) = 17%

5. Amount paid = $83.25 · 48 = $3996
   Interest charge = $3996 − $3000
   = $996
   Charge per $100 = $\frac{996}{3000}$ · 100 = $33.20
   From table:
   (48 payments and $33.59) = 15%

6. Amount paid = $26.25 · 6 = $157.50
   Interest charge = 157.50 − 150
   = $7.50
   Charge per $100 = $\frac{7.50}{150}$ · 100 = $5
   From table:
   (6 payments and $5.02) = 17%

7. From table:
   (11% interest for 30 years)
   = $9.52 per $1000
   Monthly payment = 9.52 · 80
   = $761.60

8. From table:
   ($10\frac{1}{2}$% interest for 25 years)
   = $9.44 per $1000
   Monthly payment = 9.44 · 71.5
   = $674.96

9. From table:
   (13% interest for 30 years)
   = $11.06 per $1000
   Monthly payment:
   11.06 · 122   = $1349.32
   Taxes = 960 ÷ 12 =   80.00
   Insurance = 252 ÷ 12 = 21.00
   Total   = $1450.32

10. Monthly income = $2000
    Monthly payments = $80
    $2000 − $80 = $1920
    $\frac{1920}{4}$ = $480 maximum house payment

11. Monthly income = $2800
    Monthly payments = $150
    $2800 − $150 = $2650
    $\frac{2650}{4}$ = $662.50 maximum house payment

12. From table:
    (whole life, age 25)
    = $13.36 per $1000
    Annual premium = 13.36 · 40
    = $534.40

13. From table:
    (10-year term, age 45)   *Subtract 3 from female's age*
    = $10.82 per $1000
    Annual premium = 10.82 · 55
    = $595.10

14. From table:
    (20-year endowment, age 35)
    = $35.22 per $1000
    Annual premium = 35.22 · 30
    = $1056.60

FINAL EXAM

1. 2 11
    574
    891
    3 725
    7 806
    12,996

2. $$9
    2 $\cancel{10}$ 15
    $$$\cancel{3}$ $\cancel{0}$ $\cancel{5}$9
    $-$ 2 8 74
    $$$$1 85

3. Round 28,746 to the nearest (a) ten; (b) hundred.

    (a) 28,74<u>6</u> ↑   5 or greater, 4 becomes 5. Replace ones digit with a zero.

    Answer: 28,750

    (b) 28,7<u>4</u>6 ↑   4 or less, 7 does not change. Replace ones and tens digits with zeros.

    Answer: 28,700

4. $$$$133
    74$\overline{)9859}$
    $$$\underline{74}$
    245
    $$$\underline{222}$
    239
    $$$\underline{222}$
    17

    Answer: 133 R17

5. $10\sqrt{64} - 32 \div 4$
    $= 10 \cdot 8 - 32 \div 4$
    $= 80 - 32 \div 4$
    $= 80 - 8$
    $= 72$

6. $\dfrac{125}{225} = \dfrac{5 \cdot \overset{1}{\cancel{25}}}{9 \cdot \underset{1}{\cancel{25}}} = \dfrac{5}{9}$

7. $\dfrac{5}{\underset{3}{\cancel{9}}} \cdot \dfrac{\overset{4}{\cancel{12}}}{7} = \dfrac{20}{21}$

8. $A = lw$

    $= \dfrac{2}{\underset{1}{\cancel{3}}} \cdot \dfrac{\overset{2}{\cancel{6}}}{5} = \dfrac{4}{5}$ in²

9. $\dfrac{\frac{5}{12}}{\frac{10}{3}} = \dfrac{5}{12} \div \dfrac{10}{3} = \dfrac{\overset{1}{\cancel{5}}}{\underset{4}{\cancel{12}}} \cdot \dfrac{\overset{1}{\cancel{3}}}{\underset{2}{\cancel{10}}} = \dfrac{1}{8}$

10. $\dfrac{3\frac{2}{5}}{1\frac{1}{10}} = 3\dfrac{2}{5} \div 1\dfrac{1}{10}$

    $= \dfrac{17}{5} \div \dfrac{11}{10}$

    $= \dfrac{17}{\underset{1}{\cancel{5}}} \cdot \dfrac{\overset{2}{\cancel{10}}}{11}$

    $= \dfrac{34}{11} = 3\dfrac{1}{11}$

11.
```
2 | 3   6   8   10
2 | 3   3   4   5
2 | 3   3   2   5
3 | 3   3   1   5
5 | 1   1   1   5
    1   1   1   1
```
Least common denominator
$= 2 \cdot 2 \cdot 2 \cdot 3 \cdot 5 = 120$

12. $\dfrac{3}{5} + \dfrac{1}{10} = \dfrac{2 \cdot 3}{2 \cdot 5} + \dfrac{1}{10}$

$= \dfrac{6}{10} + \dfrac{1}{10}$

$= \dfrac{7}{10}$

13. $\dfrac{5}{8} - \dfrac{1}{3} = \dfrac{3 \cdot 5}{3 \cdot 8} - \dfrac{1 \cdot 8}{3 \cdot 8}$

$= \dfrac{15}{24} - \dfrac{8}{24}$

$= \dfrac{7}{24}$

14. $\begin{array}{r} 5\dfrac{3}{4} \\ + 7\dfrac{2}{3} \\ \hline \end{array} = \begin{array}{r} 5\dfrac{9}{12} \\ 7\dfrac{8}{12} \\ \hline 12\dfrac{17}{12} \end{array} = 12 + \dfrac{17}{12}$

$= 12 + 1\dfrac{5}{12}$

$= 13\dfrac{5}{12}$

15. $8\dfrac{3}{8} = 8\dfrac{15}{40} = 7 + 1\dfrac{15}{40} = 7\dfrac{55}{40}$
    $- 2\dfrac{9}{10} = 2\dfrac{36}{40} = 2\dfrac{36}{40} = 2\dfrac{36}{40}$
    $\hspace{6em} 5\dfrac{19}{40}$

16. Round 42.0845 to the nearest
    (a) thousandth; (b) hundredth.

    (a) 42.084**5**   5 or greater,
            ↑ 4 becomes 5.
              Drop the
              ten-thousand this
              digit.

    Answer: 42.085

    (b) 42.08**4**5   4 or less, 8 does not
            ↑ change. Drop the
              thousandths and ten-
              thousandths digits.

    Answer: 42.08

17. $\begin{array}{r} 1211 \\ 5.4 \\ 19.769 \\ + 385.214 \\ \hline 410.383 \end{array}$

18. $\begin{array}{r} 99 \\ 1101010 \\ 48.2\cancel{0}\cancel{0}\cancel{0} \\ - 36.0941 \\ \hline 12.1059 \end{array}$

19. $\begin{array}{r} 6.1448 \\ 4.21_\wedge\overline{)25.87_\wedge 0000} \\ \underline{2526} \\ 610 \\ \underline{421} \\ 1890 \\ \underline{1684} \\ 2060 \\ \underline{1684} \\ 3760 \\ \underline{3368} \\ 392 \end{array}$

Answer: 6.145 (rounded to the nearest thousandth)

20. .58, $\frac{3}{5}$, $\frac{9}{16}$, .579, .5803

Write each as a decimal with 4 places.

.5800, .6000, .5625, .5790, .5803

Arrange in order, smallest to largest.

.5625, .5790, .5800, .5803, .6000

Answer: $\frac{9}{16}$, .579, .58, .5803, $\frac{3}{5}$

21. $\frac{18}{12} = \frac{\overset{1}{6} \cdot 3}{\underset{1}{6} \cdot 2} = \frac{3}{2}$

22. $\frac{5 \text{ days}}{3 \text{ weeks}} = \frac{5 \text{ days}}{21 \text{ days}} = \frac{5}{21}$

23. $\frac{55}{15} = \frac{88}{24}$

$\frac{5 \cdot 11}{5 \cdot 3} = \frac{8 \cdot 11}{8 \cdot 3}$

$\frac{11}{3} = \frac{11}{3}$

The proportion is true.

24. $\frac{3\frac{1}{2}}{9} = \frac{14}{k}$

$3\frac{1}{2} \cdot k = 9 \cdot 14$

$3\frac{1}{2} \cdot k = 126$

$\frac{3\frac{1}{2} \cdot k}{3\frac{1}{2}} = \frac{126}{3\frac{1}{2}}$

$k = 126 \div 3\frac{1}{2}$

$k = \frac{126}{1} \div \frac{7}{2}$

$k = \frac{\overset{18}{\cancel{126}}}{1} \cdot \frac{2}{\underset{1}{\cancel{7}}}$

$k = 36$

25. Let x = the unknown distance and write a proportion to solve.

$\frac{308 \text{ miles}}{11 \text{ gallons}} = \frac{x \text{ miles}}{27 \text{ gallons}}$

$11 \cdot x = 27 \cdot 308$

$11 \cdot x = 8316$

$\frac{11 \cdot x}{11} = \frac{8316}{11}$

$x = 756$

Answer: 756 miles

26. Write and solve the percent proportion.

$\frac{5}{8} = \frac{p}{100}$

$8 \cdot p = 5 \cdot 100$

$8 \cdot p = 500$

$\frac{8 \cdot p}{8} = \frac{500}{8}$

$p = 62.5$

Answer: 62.5%

27. 28% of 950

= .28 · 950

= 266

28. 48 is 15% of what number

48 = .15 · x

$\frac{48}{.15} = \frac{.15 \cdot x}{.15}$

320 = x

Answer: 320

29. $\dfrac{520}{650} = \dfrac{p}{100}$

$650 \cdot p = 520 \cdot 100$

$650 \cdot p = 52{,}000$

$\dfrac{650 \cdot p}{650} = \dfrac{52{,}000}{650}$

$p = 80$

Answer: 80%

30. The interest will be

$I = prt$

$I = 2500 \cdot .15 \cdot \dfrac{8}{12} = \$250.$

The total amount that will be due is the loan amount plus the interest.

$\$2500 + \$250 = \$2750$

31. $5 \;\cancel{\text{pints}} \cdot \dfrac{1 \text{ quart}}{2 \;\cancel{\text{pints}}} = \dfrac{5}{2} = 2\dfrac{1}{2}$ quarts

32. 5 weeks 15 days 28 hours

= 5 weeks 15 days + 1 day 4 hours

= 5 weeks 16 days 4 hours

= 5 week + 2 weeks 2 days + 4 hours

= 7 weeks 2 days 4 hours

33.   7 yards 2 feet 10 inches
    + 5 yards 2 feet 7 inches
    ─────────────────────────
     12 yards 4 feet 17 inches

= 12 yards 4 feet + 1 foot 5 inches

= 12 yards 5 feet 5 inches

= 12 yards + 1 yard 2 feet + 5 inches

= 13 yards 2 feet 5 inches

34. $F = \dfrac{9}{5}C + 32$

$F = \dfrac{9}{\cancel{5}} \cdot \overset{17}{\cancel{85}} + 32$

$= 153 + 32$

$= 185°$

Answer: 185°F

35. $A = lw$

$= 6.2 \cdot 9.4$

$= 58.28$

$= 58.3 \text{ m}^2$ (rounded)

36. $A = \dfrac{1}{2}h(b + B)$

$= \dfrac{1}{2} \cdot 4.1(5.4 + 9.8)$

$= \dfrac{1}{2} \cdot 4.1(15.2)$

$= \dfrac{1}{2} \cdot 62.32$

$= 31.16$

$= 31.2 \text{ cm}^2$ (rounded)

37. $A = \dfrac{1}{2}bh$

$= \dfrac{1}{2} \cdot 4.6 \cdot 5$

$= 2.3 \cdot 5$

$= 11.5 \text{ ft}^2$

38. $A = \pi r^2$

$= 3.14 \cdot 9.7^2$

$= 3.14 \cdot 94.09$

$= 295.4426$

$= 295.4 \text{ cm}^2$ (rounded)

39. $v = \pi r^2 h$
   $= 3.14 \cdot 5.2^2 \cdot 2.9$
   $= 3.14 \cdot 27.04 \cdot 2.9$
   $= 84.9056 \cdot 2.9$
   $= 246.22624$
   $= 246.2 \text{ cm}^3$ (rounded)

40. leg $= \sqrt{17^2 - 8^2}$
   $= \sqrt{289 - 64}$
   $= \sqrt{225}$
   $= 15$

41. $-2 + (-8) = -10$

42. $-3 - (-11) = -3 + 11 = 8$

43. $8k - 9z = 8\left(-\frac{1}{2}\right) - 9(-3)$
   $= -4 - 9(-3)$
   $= -4 - (-27)$
   $= -4 + 27$
   $= 23$

44. $3p + 1 = 19$
   $3p + 1 - 1 = 19 - 1$
   $3p = 18$
   $\frac{3p}{3} = \frac{18}{3}$
   $p = 6$

45. $8r - 9 = 12r - 1$
   $8r - 9 + 9 = 12r - 1 + 9$
   $8r = 12r + 8$
   $8r - 12r = 12r - 12r + 8$
   $-4r = 8$
   $\frac{-4r}{-4} = \frac{8}{-4}$
   $r = -2$

46. 13, 18, 19, 22, 26, 26, 28, 37, 64

   The means is
   $$\frac{13 + 18 + 19 + 22 + 26 + 26 + 28 + 37 + 64}{9}$$
   $= \frac{253}{9} = 28.1$ (rounded).

   The median is 26. (Four numbers are below it and four are above it.)
   The mode is 26. (It is the only number that is listed more than once.)

47. 4.2, 5.9, 6.3, 7.8, 8.0, 10.4, 12.5, 12.5, 24.2

   The mean is
   $$\frac{4.2+5.9+6.3+7.8+8.0+10.4+12.5+12.5+24.2}{9}$$
   $= \frac{91.8}{9} = 10.2.$

   The median is 8.0. (Half the numbers are below it and half are above it.)
   The mode is 12.5. (It appears in the list more frequently than any other number.)

48. From table:
   $(5 \cdot 2 = 10, \ 8\% \div 2 = 4\%)$
   1.4802
   Compound amount $= 1.4802 \cdot 10,500$
   $= \$15,542.10$

49. From table:
   $(3 \cdot 4 = 12, \ 12\% \div 4 = 3\%)$
   1.4258
   Compound amount $= 1.4258 \cdot 15,000$
   $= \$21,387$

50. From table:

(12%, 30) = $10.29 per $1000

monthly payment:

$10.29 \cdot 128$ = $1317.12

taxes = 1152 ÷ 12 =   96.00

insurance = 384 ÷ 12 =   32.00

total $1445.12

APPENDIX A: CALCULATORS

3. 44,904.75

7. 56.511 = 56.51 (rounded)

11. 58.90

15. 133.13772 = 133.14 (rounded)

19. $\dfrac{98{,}553.6}{174.08}$ = 566.1397 = 566.14 (rounded)

23. $\dfrac{2{,}086{,}000}{360}$ = 5794.4444

    = 5794.44 (rounded)

27. .007292 × $42,798.46

    = $312.08637

    = $312.09 (rounded)

APPENDIX B: INDUCTIVE AND DEDUCTIVE REASONING

3. 1 + 5 = 6
    6 + 5 = 11
    11 + 5 = 16
    16 + 5 = 21

    The next number will be 21 + 5 = 26.

7. 1 × 3 = 3
    3 × 3 = 9
    9 × 3 = 27
    27 × 3 = 81

    The next number will be 81 × 3 = 243.

11. See text for answer.
    The horizontal segment moves from middle to bottom to top, first all on the right of the vertical segment, then all on the left of the vertical segment.

15. The premises are

    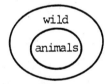 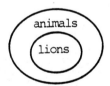

    Put these together to obtain

    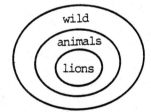

    From this diagram it may be seen that all lions are wild, and so the conclusion follows.

19. Use a Venn diagram.

    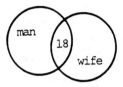

    The 18 days they watched television together are indicated above. The man watched it 20 days and 20 − 18 = 2:

    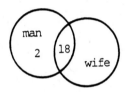

    His wife watched it 25 days and 25 − 18 = 7:

    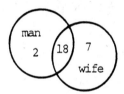

    This accounts for 2 + 18 + 7 = 27 days when one or both watched television, so on 30 − 27 = 3 days neither watched television.

NOTES

NOTES

NOTES

NOTES

NOTES